TOXIC PLANTS

Proceedings of the 18th Annual Meeting of the Society for Economic
Botany, Symposium on "Toxic Plants", June 11–15, 1977.

The University of Miami, Coral Gables, Florida

11/10/81

Symposium on Toxic Plants

→TOXIC PLANTS

A. Douglas Kinghorn, Editor

NEW YORK   COLUMBIA UNIVERSITY PRESS   1979

Library of Congress Cataloging in Publication Data

Symposium on Toxic Plants, University of Miami, 1977.
  Toxic plants.

  Includes bibliographies and index.
  1. Poisonous plants—Toxicology—Congresses.
2. Poisonous plants—Congresses.   I. Kinghorn,
A. Douglas.   II. Society for Economic Botany.
III. Title.
RA1250.S98 1977   615.9'52   79-16180
ISBN 0-231-04686-3

Columbia University Press
New York   Guildford, Surrey

Contents

Preface   vii

Contributors   xi

1. The Problem of Poisonous Plants. *John M. Kingsbury*   1

2. Toxic Mushrooms. *George M. Hatfield*   7

3. Toxins and Teratogens of the Solanaceae and Liliaceae. *Richard F. Keeler*   59

4. Pokeweed and Other Lymphocyte Mitogens. *Alexander McPherson*   83

5. Literature Review and Clinical Management of Household Ornamental Plants Potentially Toxic to Humans. *Ara Der Marderosian and Frank C. Roia, Jr.*   103

6. Cocarcinogenic Irritant Euphorbiaceae. *A. Douglas Kinghorn*   137

7. The Poisonous Anacardiaceae. *Harold Baer*   161

8. Contact Hypersensitivity and Photodermatitis Evoked by Compositae. *G. H. Neil Towers*   171

Index   185

Preface

The dependency of the human race on plants as renewable sources of many foodstuffs, drugs, textiles, fuel, and building materials is generally acknowledged. However, the harmful effects produced by certain plant constituents, which result in an increasing incidence of poisoning cases in the United States each year, are not so widely appreciated. This book is not intended to be a comprehensive treatise on all substances from plants that are lethal or otherwise injurious to humans or livestock. Instead, owing to tremendous recent advances in the understanding of the chemical nature of their toxic principles, it reviews selected toxic plants which have been hitherto inadequately documented in previous texts.

The eight chapters of this book were originally presented as a symposium at the 18th Annual Meeting of the Society for Economic Botany, held at Coral Gables, Florida, in 1977. All the contributors are active researchers in the subject areas they describe, and several of the chapters have been updated since the symposium. Four main aspects of the study of toxic plants are discussed in the book. These are the problems affecting the generation of accurate scientific and clinical data on poisonous plants, recent studies on some plants with lethal and severe toxic effects, an objective review of harmful domestic plants, and accounts of dermatitis-producing plants which cause humans a great deal of suffering.

It is hoped that the book will be of interest to both the specialist and nonspecialist reader. It should be of use to agronomists, botanists, chemists, horticulturalists, pharmacognosists, and health professionals, including physicians, pharmacists, and veterinarians, as well as the staffs of poison control centers. Equally, enthusiasts in gardening, floriculture, and those with a concern for threats to human health in the environment should find this book informative.

In the first paper, John M. Kingsbury, a widely acknowledged writer on North American poisonous plants, provides a stimulating and provocative look at the uneven quality of the sources of information in this area. He also probes the possible consequences of increased legislation on the sale of domestic ornamental plants.

George Hatfield discusses poisonous mushrooms in the first of three papers on drastic poisons from plants. He reviews some recent advances on the structure, toxicity, mechanism of action, and methods of treatment of mushroom toxins. Richard F. Keeler provides an overview of the economic consequences when livestock accidentally graze on forage containing poisonous plants. He cites examples of toxins occurring in plants of the lily and nightshade families, including those which produce deformities in livestock offspring. Since extracts of tobacco leaves and blighted potato seedlings have shown teratogenic effects in laboratory animals, they are possible human health hazards.

Certain seed proteins are the most toxic plant constituents known. For example, the rosary pea is a colorful tropical seed used to make commercially available necklaces in the United States; it contains abrin, a protein similar to cholera, tetanus, and diphtheria bacterial toxins. Alexander McPherson describes methods to isolate and characterize complex proteinaceous substances from pokeweed and other plants.

Ara Der Marderosian and Frank Roia Jr. have attempted to eliminate some of the misapprehensions concerning the safety of household plants in a chapter which combines their laboratory screening results with an extensive literature review. An important section of their chapter is a discussion of clinical practices used in the management of poisoning cases involving plants.

The final three papers focus on plants which are toxic externally to humans. A. Douglas Kinghorn discusses the irritant compounds to the eyes and skin found in the spurge family, which occur in species found in the home, garden, and commerce. Some of these are particularly significant because they have been shown to promote skin cancer in mice. Harold Baer has been involved in the most extensive study to date on plants such as poison ivy and poison oak, which cause much human suffering in the United States. He reviews the chemical nature of the active principles and immunological studies done on them, including those with human volunteers. Neil Towers describes agents from the composite family that generate allergic contact dermatitis. One plant in this family, *Parthenium hysterophorus*, is curious in that it has caused severe dermatitis in India, though not in other countries where it grows.

I would like to thank the Coca-Cola Company, Atlanta, GA: Burrough's Wellcome Company, Research Triangle Park, N.C., and Smith Kline and French Laboratories, Philadelphia, Pa., for generous financial assistance.

The authors of each chapter are to be thanked for the time and effort they have spent in preparing manuscripts. Special thanks are also due to Dr. Julia Morton, the local organizer of the meeting, for providing cordial Floridian hospitality, and to my colleagues at the Department of Pharmacognosy and Pharmacology, University of Illinois at the Medical Center, Drs. Norman Farnsworth, Harry Fong, and Edward Mika, for their interest and helpful suggestions throughout all stages of this project.

Contributors

Harold Baer, Allergenic Products Branch, Division of Bacterial Products, Bureau of Biologics, Food and Drug Administration, Bethesda, Maryland.

Ara Der Marderosian, Department of Biological Sciences, Philadelphia College of Pharmacy and Science, Philadelphia, Pennsylvania.

George M. Hatfield, College of Pharmacy, University of Michigan, Ann Arbor, Michigan.

Richard F. Keeler, Poisonous Plants Laboratory, Agricultural Research Service, U.S. Department of Agriculture, Logan, Utah.

A. Douglas Kinghorn, Department of Pharmacognosy and Pharmacology, College of Pharmacy, University of Illinois at the Medical Center, Chicago, Illinois.

John M. Kingsbury, Department of Clinical Sciences, New York State College of Veterinary Medicine, Cornell University, Ithaca, New York.

Alexander McPherson, Department of Biological Chemistry, The Milton S. Hersey Medical Center, The Pennsylvania State University, University Park, Pennsylvania.

Frank C. Roia, Jr., Department of Biological Sciences, Philadelphia College of Pharmacy and Science, Philadelphia, Pennsylvania.

G. H. Neil Towers, Department of Botany, University of British Columbia, Vancouver, B.C., Canada.

TOXIC PLANTS

I The Problem of Poisonous Plants

John M. Kingsbury

The major human problems associated with poisonous plants derive from the basic problem of a confused, seriously inadequate, and often misunderstood body of literature on the subject.

Plants produce and present toxicity in a multitude of complex ways (Kingsbury 1964; Clarke and Clark 1967; Lampe and Fagerström 1968; Morton 1971; Hardin and Arena 1974). Although vertebrates have evolved an array of mechanical and biochemical defenses against these toxins, few systems of the vertebrate body are immune to damage by some toxic compound from some plant source (Kingsbury 1975). Toxicity, by and large, involves an interrelationship among dose, absorption, detoxification, and excretion. At present, the "layman" might believe that toxicity involves the plant alone.

The word *problem* is anthropocentric, implying a human point of view. From that vantage point the problem of poisonous plants is not their direct toxicity to man alone but also the inconvenience and economic loss associated with the poisoning of domestic animals and the cost of preventing or reducing such happenings. Quality and accessibility of information is central to effective management of all these problems. This chapter examines five specific topics related to such information.

Plants with Known Toxic Constituents Are Not Always Poisonous

Primary compounds can be defined as those required for a plant's basic metabolism. Secondary compounds, loosely, are all others. In the past twenty years, knowledge of the identity, complexity, and quantity of secondary compounds in plants has increased geometrically. Patterns are beginning to emerge from the masses of data that phytochemists have produced. One of these relates to toxicity.

In living organisms probably no metabolic cycle is entirely complete or completely balanced. Secondary compounds perhaps first appeared in evolution as end products in incomplete cycles. Many of these compounds would be toxic to the organisms producing them if allowed to accumulate or become concentrated in living tissues. When plants made the landward migration hundreds of millions of years ago, they lost the ability to dissipate water soluble and lipid soluble wastes directly into the ambient environment. Land plants evolved self-protective ways of handling potentially toxic secondary compounds by removing them physically into metabolically inactive locations (e.g., bark) or by converting them into nontoxic compounds via specific chemical reactions. At the same time, plants found that secondary compounds gave them an opportunity for effective defense against attack by herbivorous insects. Coevolution of complex patterns of insect attack and plant defense involving secondary compounds leads to the conclusion that, whatever the reason for their initial appearance on the evolutionary scene, toxic secondary compounds are now fundamental to the success of plants in defending themselves, and that their defense by this means is mainly against the pervasive pressures of herbivorous insects. That many of these secondary compounds are also toxic to herbivorous vertebrates seems almost accidental as far as the selective value to plants is concerned (Kingsbury, 1978).

J. A. Duke (personal communication) has recently summarized a vast amount of information about plants and toxic secondary compounds in two tables. The first lists chemicals known to occur in plants that are also given in the 1975 *Registry of Toxic Effects of Chemical Substances.* Duke's second table lists the plants containing these compounds. These lists name about 750 toxic compounds in more than 1,000 species of plants.

One might infer from these lists that all the plants named are potentially toxic. In order for a plant to be functionally poisonous, however, it must not only contain a toxic secondary compound but also possess effective means of presenting that compound to an animal in sufficient concentration, and the secondary compound must be capable of overcoming whatever physiological or biochemical defenses the animal may possess against it. Thus, the presence of a known poisonous principle, even in toxicologically significant amounts, in a plant does not automatically mean that either man or a given species of animal will ever be effectively poisoned by that plant. Plants should not be listed as poisonous unless they have been reported to have poisoned a person or animal, or can reasonably be shown to satisfy the other requirements just discussed.

The Uneven Quality of Existing Information

In this context reliable information about actual poisonings becomes especially important. We look to the scientific literature for accurately defined cases of poisoning or, better yet, experimental confirmation of poisonings. The quality of the scientific literature as a whole comes immediately into question. It is spotty. Some sources are modern and the work excellent. Others are ancient, and that fact is often obscured by the tendency of authors to retain what appears to be useful information even when they are not immediately able to identify its origin. For example, the source of information about the toxicity of *Daphne mezereum* as it appears in a contemporary compendium of poisonous plants can be traced back (Kingsbury, 1961), reference by reference, to experiments with dogs conducted by M. J. B. Orfila before 1800. Even though his work was experimental (not always the case in those days), Orfila added little to the literature that had not already been set forth by Dioscorides nearly two millennia earlier. This is not an isolated example. A significant amount of information in contemporary compendia is derived innocently from ancient sources or observations. Persons using such compendia have an obligation to recognize and evaluate for themselves the quality of the individual pieces of information collected therein.

Access to Information Requires Preliminary Accurate Identification

An important contemporary source of information about plant poisoning in man is in the records of poison control centers as collected and analyzed by the National Clearinghouse for Poison Control Centers. Also, these inform us about the use of information by those dealing with human ingestion of toxins. Since the advent of the "childproof" safety cap on bottles of prescribed medicines and the consequent rapid decline in the incidence of poisoning by aspirin, poisonous plants have moved into first place as a reported category in the annual Clearinghouse reports (HEW 1976).

Since toxic plants do not bear labels, ingestion of plants represents a serious problem for poison control centers. Access to whatever scientific information may exist about a given plant, then, must first involve accurate determination of the botanical identity of the ingested material by someone at the poison control center. Rarely do such personnel have the necessary technical background to accomplish this. Reliance on a common name reported by a parent when a child has ingested a plant (the usual case) can result in treatment errors. A survey of plant ingestions reported by poison control centers brings this and collateral problems into sharp relief (Kingsbury 1969). Without an accurate plant name, access to the literature is defeated before it has begun, and tallies of ingestions by poison control centers are no better than the identifications that accompany them.

Physicians react to inadequate information by playing it safe. Few ingestions of plants result in hospitalizations, and fewer yet in mortalities. Yet many result needlessly in induced vomition, gastric lavage, a sense of frustration on the part of the physician, and a state of panic on the part of the parent. Much of this national agony could be prevented if the information base were reliable and capable of being accurately employed by personnel at poison control centers and by pediatricians in private practice.

Should Poisonous Ornamental Plants Bear Labels?

Reliable access to existing information could be obtained if plants bore accurate labels like cans of drain cleaners. Federal agencies have recently pursued that objective. They propose that the most dangerous plants in interstate commerce be labeled as hazardous. Plant nurserymen and commercial horticultural associations have resisted vigorously. While the federal agencies believe that some plants are fundamentally dangerous, horticulturalists argue that they are not dangerous as people use them. Both sides turn to a patently imperfect scientific literature to defend their positions. Neither side gives attention to the current overreaction of poison control centers. The resolution of this adversary position will come eventually, no doubt, from the courts.

Whatever the resolution, it is impossible to restrict or even to label effectively all species of plants containing potentially troublesome levels of toxic secondary compounds. As pointed out earlier, such a list might include over 1,000 species right now, with more added frequently. However, many food plants contain toxic secondary compounds that for one reason or another do not actually cause trouble in normal human ingestions. A line will have to be drawn somewhere.

It can be argued that labeling potentially dangerous plants takes American humanity farther down the road toward dependence on government for protection from cradle to grave. The greater such dependence, the more dangerous the omissions. Common sense and individual intelligent appraisal tend to fall by the wayside. Is furthering this trend wise in relation to such a pervasive element of the human environment as plants? A single, simple rule could obviate the necessity for warning labels on plant materials, a rule such as "Don't eat anything not commonly recognized as wholesome." Those who would experiment with wild plants as food should accept the burden of identifying materials accurately and learning whether they might be poisonous or not before they use them, or else accepting the consequences. Society at large should not be penalized for the stupidity of a very few.

If governmental agencies require labeling of some very dangerous ornamental plants (which may be a good idea as an educational aid), how are they to approach the problem of the other occasionally dangerous plants that are not labeled? In labeling some, does not the government lead the public to assume that the absence of such a label implies that the plant is never dangerous to the public under any circumstances? In any event, given the present state of information, determining which of our plants are truly the most dangerous is an exercise in uncertainty.

The Difficulty of Improving the Present Body of Information

Toxicity is rarely an all-or-none phenomenon. Species of plants vary in their content of toxic compounds owing to unpredictable extrinsic and genetic factors (Kingsbury 1960). Vertebrate species and individual animals vary in susceptibility. To describe adequately the toxicity of a given species of plant to man and the larger vertebrates is thus difficult without experiments involving a variety of plant materials fed to many individuals. Larger domestic vertebrates are expensive, and the facilities needed to experiment humanely and productively with such animals are neither easily nor cheaply obtained. Whereas the toxicity of insecticides and food additives can be explored with thousands of laboratory animals, establishing the LD_{50} of "nightshade" to "cattle" is a practical impossibility. Furthermore, to describe the functional toxicity of a particular plant adequately requires the professional capacities of a range of specialists, from botanists to pathologists or animal husbandmen to clinical toxicologists. Institutions at which appropriate teams of investigators can be brought together and provided with the necessary facilities are few. Thus, the experimental generation of useful new information about poisonous plants as they relate to man and his domestic vertebrates is an unusually difficult practical problem.

REFERENCES

Clarke, E. G. C. and M. L. Clark. 1967. *Garner's Veterinary Toxicology*, 3d ed. Baltimore: Williams and Wilkins.

Hardin, J. W., and J. M. Arena. 1974. *Human Poisoning from Cultivated Plants*, 2d ed. Durham, N.C.: Duke University Press.

HEW. 1976. "Tabulations of 1974 Case Reports." In *National Clearinghouse for Poison Control Centers Bulletin*. Bethesda, Md.: U.S. Department of Health, Education and Welfare.

Kingsbury, J. M. 1960. "Poisonous Plants of Particular Interest to Animal Nutritionists." *Proc. Cornell Univ. Nutr. Conf. Feed Manuf.* 1960:14–23.

—— 1961. "Knowledge of Poisonous Plants in the United States—Brief History and Conclusions." *Econ. Bot.* 15:119–30.

—— 1964. *Poisonous Plants of the United States and Canada*. Englewood Cliffs, N.J.: Prentice-Hall.

—— 1969. "Phytotoxicology. I. Major Problems Associated with Poisonous Plants." *Clin. Pharmacol. Ther.* 10:163–69.

—— 1975. "Phytotoxicology." In L. J. Casarett and J. Doull, eds., *Toxicology: The Basic Science of Poisons*, pp. 591–603. New York: Macmillan.

—— 1978. "Ecology of Poisoning." In R. F. Keeler, K. R. Van Kampen, and L. F. James, eds., *Effects of Poisonous Plants on Livestock*. New York: Academic Press.

Lampe, K. F. and R. Fagerström. 1968. *Plant Toxicity and Dermatitis*. Baltimore: Williams and Watkins.

Morton, J. F. 1971. *Plants Poisonous to People in Florida and Other Warm Areas*. Miami: Hurricane House.

2   Toxic Mushrooms

George M. Hatfield

No other group of plants has a more infamous reputation for being toxic than the mushrooms, which are fleshy, spore-bearing structures of certain ascomycetes and basidiomycetes. Some mushrooms, such as those containing the deadly amatoxins, certainly deserve this reputation, but the vast majority of the several thousand species of fungi in this group have not been reported to be toxic. Of course, not all species have been tested for toxicity in laboratory animals or through human ingestion. But from the data available only about 100 species are known to produce toxic effects when ingested, and of these fewer than 10 can rightfully be considered deadly.

Although no complete data are available regarding the occurrence of mushroom intoxication in the United States, the incidence does appear to be rather low. For 1975 the National Clearinghouse for Poison Control Centers reported over 170,000 poisoning cases due to all causes (Food and Drug Administration 1977). These cases were reported to the Clearinghouse by participating poison control centers and do not represent the total number occurring in the United States, since many poisonings are treated by private physicians and by hospitals not associated with a poison control center. Of the total, only 1,508 (0.9 percent) were due to mushroom ingestion. Only 128 cases developed symptoms, and of these, 71 required hospitalization. No deaths were reported for 1975, but on the average two or three fatal intoxications are reported each year in the United States (Buck 1964; Benedict 1972). Over 70 percent of the cases reported to the Center for 1975 were in children under the age of 5. The accidental ingestion of mushrooms by this age group tends to inflate the statistics, since most of these cases do not actually represent intoxications. Only 4 of the 1095 cases for the under-5 age group required hospitalization.

In the United States some of the most reliable statistics dealing with mushroom poisoning are for Colorado, where the Rocky Mountain Poison Center has acted as a clearinghouse for nearly all poisonings in that state.

Colorado has averaged about 20 cases per year, of which approximately one-third required hospitalization (Mitchel 1976). The incidence of mushroom poisoning can be expected to increase in the future, since the collection of wild mushrooms for food is becoming increasingly popular.

Until recently, one of the most widely used methods of categorizing toxic mushrooms was that utilized by Tyler (1963) in his comprehensive review of the subject in 1963. All poisonous species were grouped into four pharmacologic categories: protoplasmic poisons, neurotoxins, gastrointestinal irritants, and those producing disulfiram-like effects. Since Tyler's review, tremendous progress has been made toward establishing the chemical nature of the constituents of mushrooms that are responsible for their toxic effects, and a chemical classification is now possible, as shown in table 2.1. The

Table 2.1   Constituents of Toxic Mushrooms

Toxic Constituent	Mushroom
Amatoxins	*Amanita, Galerina, Conocybe,* and *Lepiota* species
Ibotenic acid, muscimol	*Amanita muscaria,* A. *pantherina,* and A. *cothurnata*
Psilocybin, psilocin	*Psilocybe, Panaeolus, Conocybe,* and *Gymnopilus* species
Muscarine	*Inocybe, Clitocybe,* and *Amanita* species
Coprine	*Coprinus atramentarius*
Gyromitrin	*Gyromitra esculenta*
Unknown	*Chlorophyllum, Cortinarius, Entoloma, Hypholoma,* and *Tricholoma* species

toxic components of a few species remain unidentified; they make up the last group in the table. With the exception of the *Cortinarius* species and *Hypholoma fasciculare* (Huds. ex Fr.) Kummer, the mushrooms listed in this group produce primarily gastrointestinal effects such as nausea, vomiting, and/or diarrhea. Little or nothing is known about the nature of the toxic constituents of these mushrooms, primarily because no suitable animal model exists to study this type of toxicity. Reviews by Tyler (1963) and Benedict (1972) can be consulted for information regarding the effects of these mushrooms.

Table 2.1 serves as an outline of the review to follow. Only species for which the active constituent(s) has been identified will be covered. Fortunately, this includes the majority of the mushrooms that have been reported to cause human intoxications. The review covers information available up to September 1977.

Amatoxins

Mushrooms Containing Amatoxins

The most toxic of all mushrooms thus far studied are those that contain the cyclopeptidic amatoxins. These mushrooms are truly deadly and are responsible for about 95 percent of all fatal cases of mushroom intoxication. The mushroom that is best known for these compounds is *Amanita phalloides* (Fr.) Secr., the infamous "grune Knollenblatterpilz" of Europe. The mushroom is usually 10 to 15 cm tall with a slightly vaulted cap up to 12 cm in diameter. The cap is smooth and more or less olive-green. This greenish color is most apparent in specimens that have been covered by leaves; the carpophore may appear almost white if exposed to sunlight during development. The lamellae (gills) are white, as are the spores. The "button" stage of the carpophore is completely covered by a membrane, which breaks open as the stock (stipe) and cap expand. The remains of this membrane enclose the base of the stipe in a cuplike structure termed the *volva*. Another membrane that encloses the developing lamellae on the underside of the cap forms a ring of tissue (*annulus*) on the upper part of the stipe in the mature carpophore. The white gills and spores, as well as the presence of a volva and annulus, are important structural features in differentiating A. *phalloides* and related "deadly" amanitas from edible mushrooms that may otherwise resemble these fungi. It is interesting to note that A. *phalloides* has been reported to be delicious by those who have survived its deadly effects.

Until recently A. *phalloides* was quite rare in the United States. However, in recent years it has been found with increasing frequency, especially in California (Duffy and Vergeer 1977) and the Northeast (Tanghe and Simons 1973; Yocum and Simons 1977). It has been proposed that A. *phalloides* may have been recently introduced into the United States as a contaminant of imported plants. As with other species of *Amanita*, A. *phalloides* forms mycorrhizal associations with the rootlets of many deciduous trees.

Several other species of *Amanita* that also contain these toxins and are much more prevalent in North America are A. *bisporigera* Atk. (Tyler et al. 1966), A. *ocreata* Peck (Horgan, Ammirate, and Thiers 1976), A. *suballiaceaa* (Murr.) Murr. (Stark, Kimbrough, and Preston 1973), A. *tenuifolia* (Murr.) Murr. (Block, Stephens, and Murrill 1955), A. *verna* (Bull. ex Fr.)

Pers. ex Vitt. (Weiland, Schiefer, and Gebert 1966), and A. *virosa* Secr. (Tyler et al. 1966). Unlike A. *phalloides,* these mushrooms are usually entirely white. Although these species are difficult to distinguish by their field characteristics, they can be differentiated from edible mushrooms by the presence of the volva and white lamellae and spores.

The amatoxins have also been detected in three other genera of mushrooms: *Galerina* (Tyler and Smith 1963; Tyler et al. 1963), *Conocybe* (Brady et al. 1975), and *Lepiota* (Gerault and Girre 1975). Toxic species of *Galerina* are relatively small and unappealing; however, they can be mistaken for some edible species [such as *Armillaria mellea* (Fr.) Quelet.], and several intoxications involving these mushrooms have been reported in the United States (Tyler and Smith 1963; Tyler et al. 1963). Human intoxications due to the amatoxin-containing species of *Conocybe* and *Lepiota* have not been reported from North America (Benedict 1972).

Symptoms of Intoxication

The symptoms of amatoxin poisoning have a delayed onset of about 12 (6–24) hours. The initial symptoms consist of severe abdominal pain, violent vomiting, and a cholera-like diarrhea. These effects usually cause the individual to seek medical attention; and if appropriate therapy is administered to correct the loss of electrolytes and fluid, the patient usually survives this phase of the intoxication. The gastrointestinal effects of the toxins usually subside within 24 hours. The second phase of the intoxication is a period during which the patient is relatively symptom-free. This remission is temporary, however, and usually lasts only a day or so. The third and most serious phase then develops. This phase involves symptoms arising from the liver and kidney damage induced by the toxins. These include jaundice, coagulation disorders, hypoglycemia, cardiovascular deterioration, oliguria, and hepatic coma. The principal organ affected by the amatoxins is the liver, and because of the many metabolic functions of this organ the effects of hepatic failure are many, complex, and of course very serious. If death occurs, it is usually within two to five days after ingestion of the fatal meal. The mortality rate from A. *phalloides* poisoning without treatment is over 50 percent. But with appropriate supportive therapy the chances of survival are quite good, perhaps 95 percent, in cases in which there is no preexisting disease that predisposes the patient to the effects of the toxins. It should be noted that the mortality rate is dependent not only on the treatment given to the patient but also on the amount of toxin ingested. The latter is always difficult to ascertain; thus, critical evaluation of the various treatment methods that have been published, which are based on experience with a small number of poisoning cases, is difficult. A number of cases of this type of intoxication that have occurred in the United States have been reported (Becker et al. 1976; Paaso and Harrison 1975; Finestone et al. 1972; Harrison et al. 1965).

Amatoxin Structure

The chemistry of the amatoxins has been studied extensively. Much of the significant work has been done by Theodore Wieland and his collaborators. Several reviews of this work have been published (Wieland and Wieland 1959, 1972; Wieland 1967, 1968). The amatoxins are bicyclic octapeptides, of which six are known (see figure 2.1). The common structural features of this group of peptides include a sulfoxide-substituted tryptophan nucleus that forms a transannular bridge, and several of the amino acids (alanine, glycine, hydroxyproline, and isoleucine). The configuration of the sulfoxide group has been found to be R, and the S sulfoxide shows no toxicity (Buku, Altmann, and Wieland 1974). The structural differences between members of the group are relatively minor; for example, α- and β-amanitin differ only in one amino acid (aspartic acid in β and asparagine in α). γ- and ϵ-amanitin have the same relationship; they differ from α- and β-amanitin in that the former peptides contain γ-hydroxyisoleucine rather than the γ,δ-dihydroxyisoleucine found in the more common α and β-amanitins. Amanin lacks the aromatic hydroxyl group, but otherwise is identical to β-amanitin. The last compound, amanullin, does not contain a hydroxylated isoleucine and is nontoxic. Thus, this alcohol function is essential for activity. Otherwise, amanullin is identical to γ-amanitin, the most toxic of all of the known amatoxins.

Seco derivatives of the amatoxins can be prepared by mild hydrolysis using trifluoroacetic acid. The site of cleavage is adjacent to the γ-hydroxy group of isoleucine, which has a neighboring-group effect and forms a lactone derivative with the carboxyl group formed in the hydrolysis. These seco derivatives are inactive. Inactive dethio derivatives can also be prepared using hydrogen and Raney nickel. By a combination of these techniques,

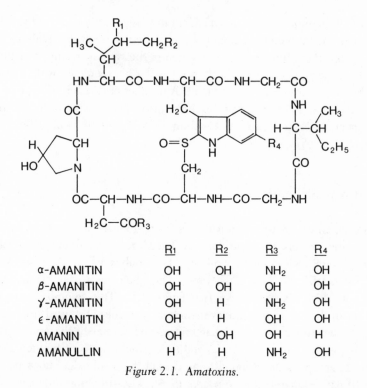

	R₁	R₂	R₃	R₄
α-AMANITIN	OH	OH	NH₂	OH
β-AMANITIN	OH	OH	OH	OH
γ-AMANITIN	OH	H	NH₂	OH
ε-AMANITIN	OH	H	OH	OH
AMANIN	OH	OH	OH	H
AMANULLIN	H	H	NH₂	OH

Figure 2.1. Amatoxins.

linear secodethio derivatives can be produced that can be sequenced by standard methods. Recently the absolute configuration of β-amanitin was established by X-ray analysis (Kostansek et al. 1977).

Assay Methods

Several assay methods have been devised for the amatoxins. Preston, Stark, and Kimbrough (1975) developed a method based on the inhibition of the enzyme RNA polymerase II, which is capable of measuring quantities down to 0.05 ng. However, this is a rather tedious method and does not distinguish between the various amatoxins. Radioimmunoassays have also been developed, but they are equally difficult because of the problems associated with producing the needed antibodies (Faulstich, Trischmann, and Zobeley 1975; Fiume et al. 1975). These assays are capable of detecting 50 pg (Faulstich, Trischmann, and Zobeley 1975) or 500 pg (Fiume et al. 1975) of amatoxins. As with the RNA polymerase II assay, these methods cannot distinguish among the different amatoxins. Faulstich, Georgopoulos, and Blocking (1973) have developed an analytical method wherein the individual amatoxins are quantified by amino acid analysis of their γ-hydroxylated leucines or isoleucines. Yocum and Simons (1977) reported recently a modification of this method in which the toxins were separated using Sephadex LH20 and quantified by their absorption at 303 nm. Finally, several TLC methods have been reported that can be used to detect toxins in suspected mushrooms, but these systems are difficult to use for quantitative analysis (Sullivan, Brady, and Tyler 1965; Raaen 1969; Palyza and Kulhanek 1970; Palyza 1974). At best, TLC can detect about 0.025 μg of amatoxin (Palyza 1974). A quantitative TLC method was recently reported that utilized reflectance and transmittance spectrometric measurement of chromatograms (Andary et al. 1977).

Toxicity

The LD_{50} for the various amatoxins varies from 0.15 to 0.5 mg/kg when administered parenterally to mice (Wieland 1972). The lethal dose in man is estimated to be approximately 0.1 mg/kg when the toxins are ingested orally. The amatoxin content of several mushroom species has been reported. The most toxic species appears to be A. *phalloides*, which contains between 0.2 and 0.3 mg of amatoxin per g of fresh mushroom tissue (Yocum and Simons 1977; Faulstich et al. 1974; Johnson, Preston, and Kimbrough 1976). Thus, it is conceivable that ingestion of one mushroom weighing about 50 g might be sufficient to kill the average individual. The other amatoxin-containing species of *Amanita* (Yocum and Simons 1977; Preston, Stark, and Kimbrough 1975; Faulstich et al. 1974) and *Galerina* (Faulstich et al. 1974; Johnson, Preston, and Kimbrough 1976) are usually less toxic and contain from 0.05 to 0.25 mg per g (fresh weight). The toxin content of A. *verna* and A. *virosa* appears to be quite variable, and some specimens of the former species have been found to be devoid of the amatoxins (Yocum and Simons 1977). Recently several edible mushrooms (e.g., *Boletus edulis* and *Cantharellus cibarius*) were found to contain trace amounts of amatoxins (Faulstich and Cochet-Meilhac 1976), and it has been suggested that these cyclopeptides might have some significance in the development of basidiomycetes. The concentration in the nontoxic species assayed was found to be 1–10 ng/g of fresh tissue.

Mechanism of Action

The mechanism of action of the amatoxins has been studied in detail. Histologic examination of hepatic and renal parenchyma cells from mice killed a few hours after administration of α-amanitin showed an intact cytoplasm and evidence of damage to nuclei (Fiume and Laschi 1965). These nuclear lesions appeared very quickly, being detectable as early as one to two hours after injection of the toxin. Some of the effects observed in the nucleus after exposure to the amatoxins are chromatin condensation, indicating inhibition of transcription; breakup of nucleoli, with separation of fibrils from granules; and accumulation of interchromatic granules at the center of the nucleus (Fiume 1972). Changes in the cytoplasm also occur, but much later (48 hours), and rapidly evolve toward necrosis (Wieland and Wieland 1972).

An indication of the biochemical mechanism underlying these effects came from studies that showed that nuclear RNA content and synthesis were drastically reduced during the first 24 hours after toxin exposure, while nuclear DNA and protein content remained essentially unchanged (Fiume and Stirpe 1966; Stirpe and Fiume 1967). Several inhibitors of RNA synthesis are known, and many, like actinomycin D, act by binding to the DNA template involved in RNA synthesis. However, several studies have shown that the polymerase enzyme, rather than the template, is the site of action of the amatoxins (Jacob, Sajdel, and Munro 1970, 1971; Kedinger et al. 1970; Lindell et al. 1970; Meihlac et al. 1970; Seifart and Sekeris 1969). Several RNA polymerases have been found to be involved in the transcription of DNA in eukaryotic cells. These enzymes are distinct in their localization, their requirements, and the nature of their templates and products, and in the inhibitors to which they are susceptible (Jacob et al. 1970). Polymerase I or A is located in the nucleolus and is concerned primarily with the transcription of genes specifying ribosomal precursor RNA. Polymerase II or B is a nucleoplasmic enzyme and is thought to be responsible for transcription of most chromosomal genes. Another polymerase (III or C) is found in the cytoplasm; its function has not been fully defined. Of these enzymes, RNA polymerase B is by far the most sensitive to the amatoxins. The toxins bind to this enzyme with a very high affinity, forming a 1:1 stoichiometric complex (Chambon et al. 1970). The binding constant for the amatoxin–polymerase complex has been determined to be about 10^{-9}M, which makes the affinity of the toxin for the enzyme in the order of those reported the strongest binding between haptens and their specific antibodies (Sperti et al. 1973). Once the toxin is bound to the enzyme, the formation of phosphodiester bonds is inhibited, thus blocking RNA synthesis (Cochet-Meihlac and Chambon 1974). This leads to the condensation of euchromatin and other effects observed in nuclei exposed to the amatoxins. Recently the amatoxin-binding subunit of RNA polymerase B was identified (Brodner and Wieland 1976). The class C cytoplasmic polymerases are also inhibited, but the concentration required is about 1,000 times that which affects RNA polymerase B (Seifart, Benecke, and Juhasz 1972; Amalrac, Nicoloso and Zalta 1972). RNA polymerase A is unaffected by the toxins (Jacob, Sajdel, and Munro 1970, 1971; Kedinger et al. 1970; Lindell et al. 1970; Seifart and Sekeris 1969; Meihlac et al. 1970). This rather specific inhibition of RNA polymerase B is being used extensively to study the control mechanisms of transcription.

The direct inhibition of RNA polymerase B leads to many indirect effects, and eventually the formation of ribosomal RNA, transfer RNA, protein, and DNA are all inhibited by the toxins (Hadjiolov, Dabeva, and Meckedonski 1974). The inhibition of RNA polymerase B has been correlated with *in vivo* toxicity, and the results of this study show a parallel between the two (Buku et al. 1971).

The principal result of these inhibitory effects is damage to hepatocytes, with partial to complete loss of hepatic function. As mentioned previously, this leads to a host of life-threatening effects. In severe cases there is a marked cytolysis of hepatocytes and terminally the liver softens, becomes painful to palpation, and atrophies. Blood levels of several enzymes that are released by damaged cells have been used to assess hepatocellular damage. SGOT, SGPT, and LDH levels are often extremely elevated in this type of intoxication.

The toxins also cause damage to the intestinal tract, and this is thought to be due in large part to enterohepatic cycling of the toxin. Studies in dogs have shown that the toxins are excreted in the bile, which leads to cellular damage in the duodenum (Fauser and Faulstich 1973). Much of the toxin excreted in this way is reabsorbed into the liver with the bile. Interruption of this cycling of the toxin by bile duct cannulation can reduce the toxic effects of α-amanitin in dogs (Fauser and Faulstich 1973). Such treatment would be unlikely to help humans poisoned by these toxins because of their short half-life in the body (*vide infra*). Oral doses of activated charcoal might be useful in removing any toxin still being excreted by the liver (Seeger and Bartels 1976). Studies in dogs have also shown that the amatoxin-induced lesions in the intestinal tract are responsible for the vomiting and diarrhea seen in the first phase of the intoxication (Fiume et al. 1973).

The kidneys are also damaged by the amatoxins, but usually to a lesser degree than the liver. In most animals, including man, the amatoxins are excreted through the glomerulus. However, some of the toxin is reabsorbed by cells of the proximal tubules, and this leads to renal tubular necrosis (Fiume and Laschi 1965). The fact that reabsorption causes the kidney damage was demonstrated using methyl α-amanitin covalently bound to albumin. The conjugate that does not pass through the mouse glomeruli did not induce kidney damage but was hepatotoxic (Fiume, Marinozzi, and Nardi 1969; Cessi and Fiume 1969; Bonetti, Derenzini, and Fiume 1976). No lesions are produced in the rat kidney after treatment with unconjugated α-amanitin, and this has been shown to be due to the incapacity of epithelial cells in the rat kidney tubules to reabsorb the toxin from the preurine (Fiume, Marinozzi, and Nardi, 1969). Since renal damage is produced in man by the amatoxins, tubular reabsorption probably occurs in the human kidney. The clinical course of a patient presenting primarily prolonged renal involvement has been reported (Myler, Lee, and Hooper 1964).

Excretion

Although some tubular reabsorption of the toxins does occur in most animals studied, the principal excretion route of these compounds is still via the kidneys. Faulstich and Fauser (1973) found that in dogs 85 percent of the administered dose of ^{14}C-methyl-α-amanitin was in the urine after 6 hours. This indicates that the toxin has a half-life of about 2.2 hours. It is difficult to say what the excretion rate of the amatoxins is in humans, but if it is in the same range as that of the dog, it is probable that a significant portion of the toxins has been excreted prior to the onset of symptoms. This is the case in the dog, which experiences vomiting and diarrhea 9 to 11 hours after toxin administration and death in 20 to 24 hours (untreated) (Fauser and Faulstich 1973; Fiume et al. 1973; Faulstich and Fauser 1973).

The excretion of α-amanitin has also been studied in mice using a radio-immunoassay for the toxin in serum (Palyza and Kulhanek 1970). The animals received 350 ng of the toxin per g of body weight (i.p.) and exhibited serum levels after an hour of approximately 100 ng/ml. The serum level dropped very rapidly, and the toxin could not be detected after 4 hours (limit of sensitivity: 0.5 ng/ml). Thus, it appears that once the toxins are absorbed they act rapidly to produce cellular damage and are then rather quickly excreted by the kidneys. The significance of the rapid excretion of the toxins, if it occurs in man, is that therapeutic measures directed toward removal of the toxins from the patient after the onset of symptoms (12 hours after ingestion) may be futile. One such mode of therapy that has been utilized is the administration of large doses of drugs such a penicillin G, chloramphenicol, or phenylbutazone to increase the excretion of the amatoxins (Moroni et al. 1976). Floersheim and associates (Floersheim 1971, 1972a; Floersheim, Schneeberger, and Bucher 1971) have found that these compounds, as well as several others, increase the survival of animals treated with lethal doses of α-amanitin if administered 30 minutes to 2 hours after the toxin. The common feature of these compounds is that they are all bound to serum protein, and thus it was proposed that they acted by displacing the amatoxins from albumin and in doing so increased the rate of elimination. However, recently it was shown that ^{3}H-methyl-demethyl-γ-amanitin does not bind to human serum albumin (Fiume et al. 1977), and there is no reason to assume that the naturally occurring toxins, such as α- and β-amanitin, have a great affinity for serum protein.

Treatment

The treatment to be given a patient with this type of intoxication should be similar to that for patients suffering from acute hepatitis due to other causes. Blood chemistry values (SGOT, SGPT, LDH, alkaline phosphatase, prothrombin time, glucose) should be followed to determine the extent and progression of liver dysfunction. Hypovolemia and electrolyte disturbances should be corrected as needed. Blood glucose levels must be maintained by orally or parenterally administered carbohydrate. Dietary protein intake should be minimized and the bowel sterilized, using an antibiotic such as neomycin, to prevent excessive blood ammonia levels, which can lead to hepatic encephalopathy. Recently hemoperfusion through charcoal treated with a biocompatible polymer has been found to be useful in treating human amatoxin poisoning (Bartels 1976). This method has been used to treat liver failure due to other causes and probably acts by removing metabolities such as phenylalanine, tyrosine, and methionine, which are thought to be involved in the pathogenesis of hepatic encephalopathy (Gazzard et al. 1974). Other methods of hemodialysis and peritoneal dialysis have been shown to be ineffectual in treating liver damage due to causes other than the amatoxins (Stiegers and Strubelt 1973).

One of the most publicized antidotes for the amatoxins is α-lipoic acid or thioctic acid. This compound attracted considerable interest in the late 1950s as a hepatotherapeutic agent in the treatment of diverse liver diseases. Kubicka (Kubicka 1963; Kubicka and Adler 1968) was apparently the first to use α-lipoic acid to treat patients poisoned by the amatoxins, and claimed improvement in 39 of 40 patients treated. Subsequently there were several other reports concerning the value of thioctic acid from Europe (Moroni et al. 1976; Delfino 1970; Ciocatto, Delfino, and Trompeo 1970; Ciucci and Chiri 1974; Zulik, Bako, and Badavari 1972) and the United States (Becker et al. 1976; Paaso and Harrison 1975; Finestone et al. 1972). Doses of up to 500 mg/day have been used, and the only side effect appears to be hypoglycemia, which may be due in part to the hepatic damage produced by the amatoxins. It is very difficult to evaluate the effectiveness of thioctic acid from the clinical reports just cited, for a number of reasons. One is that it is impossible to know just how much of the poisonous material was actually ingested. Also, in all the cases several therapeutic measures were employed in an effort to save the patient, and this makes it impossible to determine the actual benefit of any one factor, such as thioctic acid. As with other drugs, effectiveness could certainly be tested with well-controlled animal studies. Floersheim (Floersheim 1971) showed that thioctic acid was completely without value in mice poisoned with α-amanitin or extracts of A. *phalloides*. More recently Alleva (Alleva 1975; Alleva et al. 1975) obtained similar results in both mice and dogs. But this work has been criticized because seemingly inadequate doses of glucose were administered to the dogs to reverse the hypoglycemia that developed (Bartler 1975). In order to clarify this situation, further studies will be needed, preferably using dogs, which show a sensitivity to the toxins similar to that of man. These animals are also large enough to receive the supportive treatment usually afforded man. In many respects thioctic acid is the Laetrile of mushroom intoxications.

A number of other treatment modes have been claimed to be of value.

These include organotherapy (administration of a slurry prepared from rabbit brains and stomachs) (Limousin 1959), hyperbaric oxygenation (Goulon 1967), whole-blood transfusions (Gaultier et al. 1962), sex hormone therapy (Truhaut et al. 1973), treatment with lipotropic factors (Fournier and Ripault 1965), vitamin therapy (Bastien 1973; Cavalli, Ragni, and Zuccaro 1968; Dyk, Piotriwska-Sowinska, and Buczkowska 1969), and treatment with cytochrome C (Stiegers and Strubelt 1973; Floersheim 1972b). None of these has been shown to be effective in man.

In summary, there is currently no antidote for amatoxin poisoning that is sufficiently well established by both experimental and clinical testing for unequivocal recommendation. For the present, treatment should consist of supportive therapy to counter the effects of the severe liver damage induced by the toxins.

Phallotoxins

In addition to the amatoxins, A. *phalloides* contains several other toxic constituents that have received considerable study. One such group is the phallotoxins, which consist of seven closely related bicyclic heptapeptides (Wieland and Wieland 1959; Wieland 1967, 1968; Wieland and Wieland 1972). As shown in figure 2.2, these compounds are structurally related to the amatoxins. The phallotoxins contain one less amino acid in the peptide ring (asparagine or aspartic acid), a thio ether bridge rather than one involving a sulfoxide, and no phenolic hydroxyl, which is found in most of the amatoxins. Seco and dethio derivatives of the phallotoxins are inactive, as with the amatoxins, but the γ-hydroxyl group is not essential for activity.

The phallotoxins have an LD_{50} in the mouse that ranges from 1.5 to 15 mg/kg, which makes these compounds much less toxic than the amatoxins (Preston, Stark, and Kimbrough 1975; Faulstich et al. 1975). In addition, the phallotoxins appear to be even less toxic orally. Thus, the role of the phallotoxins in A. *phalloides* intoxications is thought to be relatively minor. Another reason for assuming that the phallotoxins are not significant in delayed-onset *Amanita* poisoning is that animal studies have shown that these compounds can produce death in 1 to 4 hours from a lethal dose, in contrast to the amatoxins, which rarely cause death in less than 15 hours.

In spite of the apparent insignificance of the phallotoxins in human intoxications, their toxic effects and mechanism of action have been extensively

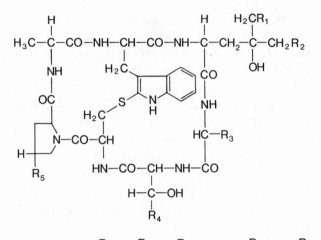

	R₁	R₂	R₃	R₄	R₅
PHALLOIDIN	-OH	-H	-CH₃	-CH₃	-OH
PHALLOIN	-H	-H	-CH₃	-CH₃	-OH
PHALLISIN	-OH	-OH	-CH₃	-CH₃	-OH
PHALLACIDIN	-OH	-H	-CH(CH₃)₂	-COOH	-OH
PHALLACIN	-H	-H	-CH(CH₃)₂	-COOH	-OH
PHALLISACIN	-OH	-OH	-CH(CH₃)₂	-COOH	-OH
PHALLIN B	-H	-H	-CH₂C₆H₅	-CH₃	-H

Figure 2.2. Phallotoxins.

studied. When phallotoxins are injected into experimental animals, their deleterious effects are confined to the liver (Siess, Wieland, and Miller 1970). The liver of treated animals becomes greatly enlarged, dark red in color, and very brittle. These changes are due to an excessive accumulation of blood in this organ. Numerous studies have shown that the phallotoxins act by binding to an actin-like protein in the cell membrane of hepatocytes (Govindan et al. 1972; Lutz, Glossman, and Frimmer 1972; Govindan et al. 1973; Lengsfeld et al. 1974; Wieland 1977). It has been proposed that the actin-like protein is located close to or within the membrane and constitutes an essential stabilizing component that is released when it combines with the toxin. Once it has been released, the integrity of the membrane is lost, leading to the loss of potassium ions and enzymes, both of which have been found to be released from phalloidin-treated livers (Frimmer and Kroker 1973). A number of studies have been directed toward the discovery of an effective antidote to the phallotoxins (Dolara, Buiatti, and Geddes 1971; Floersheim 1974; Frimmer and Petzinger 1975; Tuchweber et al.

1972). However, because the mechanism of action of the phallotoxins is completely different from that of the amatoxins, it is unlikely that this work would apply to the treatment of human intoxications due to A. *phalloides*.

Phallolysin

Another toxin of *Amanita phalloides* that appears to have no significance in human intoxications is phallolysin. This material is actually a mixture of two glycoproteins (Faulstich and Weckauf-Blocking 1974; Seeger 1975a,b). Phallolysin is not toxic when administered orally but is extremely toxic if given intravenously. Death results from intravascular hemolysis, which leads to renal failure and cardiac arrhythmias, the latter being due to an elevation in serum potassium following massive erythrocyte hemolysis (Odenthal, Seeger, and Vogel 1975). The phallolysins act by binding to N-acetylglucosamine on the erythrocyte surface (Faulstich and Weckauf 1975). These lectin-like compounds have been found in several other *Amanita* species, some of which are known to be nontoxic when eaten (Seeger, Kraus, and Wiedmann 1975).

Antamanide

Wieland and associates (1968) have isolated another peptide from A. *phalloides* that appears to decrease the toxicity of phalloidin and α-amanitin if administered prior to the toxins. This compound, which has been named antamanide, appears to delay the absorption of phalloidin into the liver (Wieland et al. 1972; Faulstich et al. 1974). Antamanide is a cyclic decapeptide containing alanine, valine, phenylalanine, and proline, and it has been prepared synthetically by several routes (Wieland et al. 1968).

Ibotenic Acid and Muscimol

Amanita muscaria Intoxication

Another species of *Amanita* that has long been known for its toxic effects is *Amanita muscaria* (Fr.) Hook. This species contains only trace amounts

of the amatoxins, and thus these constituents are not toxicologically significant (Fiume et al. 1975). The intoxication produced by A. *muscaria* is characterized by a variety of central nervous system effects. The onset of symptoms is usually after 30–60 minutes, and initially the intoxication resembles
that produced by alcohol, with drowsiness, euphoria, dizziness, and ataxia.
Nausea and vomiting may occur, but these are not consistent effects. A
delirious state then develops, with confusion, agitation, visual and auditory
disturbances, and skeletal muscle spasms. These effects usually last several
hours and are often followed by drowsiness and sleep. Recovery is usually
complete within 6–24 hours (Waser 1967). Death due to A. *muscaria* intoxication is rare, and when it does occur the victim was frequently in a precar-
ious state of health prior to the poisoning. Thus, this mushroom's reputation
for deadly toxicity is largely undeserved. It should be noted that the symptoms of A. *muscaria* intoxication are extremely variable and depend not only
on the amount of toxic material ingested but also on the set and setting (Ott
1976a).

A. *muscaria* has a long and interesting history of use as an intoxicant,
especially by certain tribes in northern Siberia and the Kamchatka Peninsula
(Wasson and Wasson 1957; Wasson 1967, 1968). These tribes had the very
unusual practice of ingesting the urine from an individual intoxicated by the
mushroom. In doing so, the person ingesting the urine reportedly developed
an intoxication similar to that produced by the mushroom. One might say
they were "passing" the intoxication from one person to another. There is
some evidence that active constituents of A. *muscaria* are, at least in part,
excreted unchanged in the urine (Waser 1967; Ott, Wheaton, and Chilton
1975).

Wasson (1968) has published a thorough review of the literature dealing
with the Siberian use of A. *muscaria* in support of his theory that this
mushroom represents the euphoriant plant Soma described in ancient Indian literature (i.e., in the Rig Veda). The identity of Soma, which is no
longer used in India as an intoxicant, has long been an ethnobotanical mys-
tery. One of the principal points favoring Wasson's proposal is that the Rig
Veda seems to describe Soma both as a plant and as the urine of priests who
have ingested the plant (Wasson 1968). If A. *muscaria* is in fact Soma, it
would make this mushroom one of the oldest intoxicants known to man,
dating back approximately 4,000 years.

An even more controversial theory regarding the ancient use of A. *muscaria* was proposed by Allegro (1970). He theorized that the New Testament

was written in a code designed to conceal the activities of a cult involved in the use of A. *muscaria* as an intoxicant. The proposal is based largely on linguistic evidence, and, needless to say, this theory has few supporters.

The writings of Wasson, Allegro, and others have increased the public awareness of the psychotropic effects of A. *muscaria*, and its use as a recreational drug is increasing (Ott 1976a). Indeed, many of the intoxications due to this mushroom result from intentional ingestion. Accidental poisonings are fairly rare owing to the unique and striking appearance of this mushroom. As with other Amanitas, the gills and spores are white, as is the stipe, which is somewhat enlarged at the base. The most characteristic feature of A. *muscaria* is the blood-red to yellowish cap (depending on variety), which has pieces of the universal veil adorning its surface in the form of whitish warts. Because of its unusual pharmacologic effects and appearance, this mushroom has become the stereotype image of a "toadstool" in our society. No other species of mushroom is used more frequently in various art forms to represent the "poisonous mushroom." Thus, there are few individuals in our generally mycophobic population that would be unaware that this mushroom is toxic. However, it should be mentioned that weathering of a carpophore can result in removal of the characteristic pileal warts and in bleaching of the cap color. These circumstances create opportunities for misidentification by the careless collector. The gross resemblance of the young "button" stage of this *Amanita* (and others) to the edible puffballs presents further opportunities for error.

Active Constituents

Attempts to isolate the active constituents of A. *muscaria* began over 150 years ago. Muscarine (see figure 2.3) was the first active constituent to be isolated, and for many years this compound was assumed to be responsible

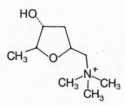

Figure 2.3. Muscarine.

for the toxic effects of the mushroom. However, this is not the case. As will be described in more detail later, muscarine is a potent parasympathomimetic and produces effects, such as sweating, nausea, vomiting, abdominal pain, blurred vision, salivation, and diarrhea, that are due entirely to peripheral stimulation of the parasympathetic nervous system. Muscarine does not affect the central nervous system because it does not pass the blood–brain barrier (Waser 1961, 1967). This compound is also poorly absorbed when taken orally, thus decreasing its toxicity by this route. In the monkey, for example, 0.5 mg given intraperitoneally produced severe toxic effects while 2 mg of muscarine by mouth produced no signs of poisoning (Fraser 1957). Since A. *muscaria* usually contains about 0.0002–0.0003 percent muscarine (fresh weight), several kg would normally have to be ingested to produce signs of muscarine intoxication. In most cases 100 g or less of the fresh mushroom will produce the CNS effects previously described. Occasionally cases of A. *muscaria* intoxication appear in which sweating and salivation are added to the usual symptoms (Waser 1967; Ott 1976a). It has been proposed that some strains of the mushroom contain higher levels of muscarine than others, but this has not been verified by analysis.

Many older references dealing with the treatment of mushroom poisoning recommend that atropine (an anticholinergic) be used to counter the effects of muscarine in A. *muscaria* poisoning. This is poor advice, for two reasons. First, although atropine is a specific antidote for this compound, muscarine intoxication is rarely a problem with this mushroom. Second, the CNS effects of atropine potentiate some of the effects of A. *muscaria*.

The fact that A. *muscaria* has some atropine-like effects has long been recognized. In 1891 Kobert proposed that the mushroom contains a constituent that blocks some of the effects of muscarine (Kobert 1891). He called this substance pilzatropine. The early literature dealing with pilzatropine has been reviewed by Tyler (1958). Reports that hyoscyamine (Lewis 1955), atropine, and scopolamine occur in A. *muscaria* could not be confirmed in subsequent studies (Brady and Tyler 1958; Salemink et al. 1963; Talbot and Vining 1963; Tyler and Groger 1964), and thus it is doubtful that these alkaloids play any role in the intoxicating effects of this mushroom.

Very little progress was made toward the identification of the compounds responsible for the CNS effects of A. *muscaria* until the mid-1960s. Then three groups working independently in Japan (Takemoto, Nakajima, and Yokoke 1964; Takemoto, Nakajima, and Sakuma 1964), England (Bowden, Drysdale, and Mogel 1965; Bowden and Drysdale 1965), and Switzerland

(Muller and Eugster 1965; Good, Muller, and Eugster 1965; Eugster, Muller, and Good 1965; Eugster 1968) reported the isolation of the two isoxazole derivatives, ibotenic acid and muscimol (see figure 2.4). The English and Japanese investigators discovered these compounds while studying the toxic effects of A. *muscaria* on flies. This mushroom is also called the fly agaric; it has been known to be insecticidal since the thirteenth century. Both ibotenic acid and muscimol are toxic to the common housefly, and this effect was used as a bioassay to aid in the isolation of these compounds.

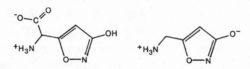

Figure 2.4. Isoxazole derivatives from Amanita muscaria.

In contrast, the Swiss investigators, led by Conrad Eugster, were attempting to isolate the constituents responsible for the CNS effects of the mushroom. They employed a narcosis-potentiating assay in mice as a guide for fractionation. These studies also led to the isolation of ibotenic acid and muscimol, which are now recognized as the principal pharmacologically active compounds of the mushroom. Since their discovery both compounds have been synthesized by several routes (Gagneux et al. 1965a, b; Krogsgaard-Larson and Christensen 1976), and a number of analogs of muscimol have also been prepared (Loev, Wilson, and Goodman 1970; Krogsgaard-Larson et al. 1975; Krogsgaard-Larson and Johnston 1975; Hjeds and Krogsgaard-Larson 1976; Lykkeberg and Krogsgaard-Larson 1976).

Pharmacologic Effects of Ibotenic Acid and Muscimol

The pharmacologic properties of ibotenic acid and muscimol have been studied, and their effects on the intact animal are similar (Waser 1967; Theobald et al. 1968). Waser (1967) reported that self-administration of 10–15 mg of muscimol produced a toxic psychosis with confusion, dysarthria, disturbance of visual perception and hearing, illusions of color vision, muscle twitching, disorientation in situation and time, weariness, fatigue, and sleep with dreams. The effects simulate a type of acute exogenic reaction but do not lead to true hallucinations such as those produced by

LSD. The LD_{50} for muscimol in rats ranges from 4.5 mg/kg i.v. to 45 mg/kg (oral) (Theobald et al. 1968). Waser also reported that a 20-mg dose of ibotenic acid produced no psychic stimulation but only lassitude followed by sleep. When taken orally, muscimol is 5–10 times as potent as ibotenic acid in man. Ibotenic acid is quite easily decarboxylated to muscimol, and it is possible that the activity of the acid is due to its partial conversion to muscimol *in vivo*. Since A. *muscaria* usually contains about equal quantities of ibotenic acid and muscimol, the latter compound must be considered the principal active constituent. However, it has been suggested that muscimol is not a genuine constituent of living A. *muscaria* and that it is produced mainly during drying and extraction by decomposition of ibotenic acid (Eugster 1968). Carpophores of A. *muscaria* vary considerably in their isoxazole content, which ranges from 0.17 to 1.0 percent (dry weight) (Benedict, Tyler, and Brady 1966). The concentration of these compounds decreases with time in dried carpophores. If it is assumed that the total isoxazole content is 0.2 percent of the dried mushroom, that ibotenic acid and muscimol are equally present in the isoxazole fraction, that ibotenic acid is one-fifth as active as muscimol, and that fresh carpophores contain 10 percent nonvolatile solids, 100 g of fresh mushroom should contain the pharmacologic equivalent of 12 mg of muscimol and would produce an intoxication if ingested.

The known pharmacological properties of ibotenic acid and muscimol do not explain all of the effects observed in A. *muscaria* intoxication, but they do clarify many of the significant clinical symptoms, such as the muscle spasms, the state of confusion and incoordination, disturbances in perception, and drowsiness followed by sleep. Several investigators have stated that other CNS-active compounds are probably present in this species, but during the past decade no new compounds have been reported that have been shown to contribute significantly to the oral toxicity of A. *muscaria*.

Muscimol is an analog of γ-aminobutyric acid, an inhibitory synaptic transmitter in the mammalian central nervous system (Roberts 1974; Brehm, Hjeds, and Krogsgaard-Larsen 1972). Johnston (1971) has shown that muscimol is a weak, noncompetitive inhibitor of the uptake of γ-aminobutyric acid by slices of rat cerebral cortex. More recently, muscimol was shown to have a high affinity for γ-aminobutyric acid receptors (Enna and Snyder 1976). When muscimol acts on γ-aminobutyric acid receptors in the cerebellum, the inhibitory function of Purkinje cells is reduced (Biggio et al. 1977). These cells are the pivotal neurons of the cerebellar cortex, and the rest of the cortical control mechanism serves to control their activity. When the inhibitory function of these cells is blocked by means other than muscimol, muscle tremors and incoordination are produced. These effects are similar to those seen in A. *muscaria* intoxication. Muscimol has become an important research tool as a γ-aminobutyric acid agonist in studying the neurochemistry of the CNS. Numerous investigations concerning the effect of muscimol on various neurons have been published in recent years.

Amanita pantherina

A number of other species of *Amanita* have been screened for the presence of ibotenic acid and muscimol (Benedict, Tyler, and Brady 1966; Chilton and Ott 1976). The only other species that have been found to constantly contain these compounds are A. *cothurnata* Atk. and A. *pantherina* (Fr.) Secr. These species appear to contain the isoxazoles in levels approximating those found in A. *muscaria*. Intoxications due to A. *pantherina* are quite common in central Europe (Pfefferkorn and Kirsten 1967) and the Pacific Northwest (Ott 1976a; Hatfield and Brady 1975). In most respects the intoxication produced by this mushroom is very similar to that produced by A. *muscaria*.

Treatment

Treatment of intoxication due to these mushrooms should consist of supportive therapy and very conservative drug use. As mentioned previously, the use of atropine (or scopolamine) is contraindicated. The use of diazepam and phenobarbital should also be avoided since there is some evidence that these drugs may potentiate the effects of muscimol (Lykkeberg and Krogsgaard-Larson 1976). The delirium or manic phase of the intoxication can usually be managed without drug intervention, utilizing reassurance and restraint as required. Since convulsive activity has been reported (especially in children), the use of stimulants during the depressive phase is also contraindicated. Despite the apparent severity of these intoxications, they are usually short-lived, rarely persisting more than 8 hours, and the chances of complete recovery are excellent.

Psilocybin and Psilocin

Another group of mushrooms that affect the central nervous system are
those that contain psilocybin and psilocin (see figure 2.5). These indole
derivatives produce LSD-like hallucinations, and the effects of mushrooms
containing these substances do not differ significantly from those observed
for the pure compounds (Cerletti and Hofmann 1963; Hollister 1961; Hollister and Hartman 1962; Wolbach, Miner, and Isbel 1962). When these

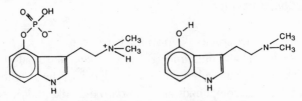

Figure 2.5. Hallucinogenic indole derivatives.

fungi are ingested, the onset of effects is usually within 30 to 60 minutes,
starting with difficulty in focusing, tinted (amber, purple, or green) vision,
and brightened colors. During this period a euphoric and/or introspective
state develops. The visual effects usually intensify to give colored patterns
and shapes, mostly with the eyes closed. Surfaces seem to have a wavy mo-
tion, and the perception of space and time is often affected. These effects
usually last 2 to 4 hours and are often considered pleasant. However, this
depends on the individual's personality, expectations, and previous experience with similar drugs. The usual oral dose needed to produce these effects
in an adult is 4 to 8 mg of psilocybin. The toxicity of psilocybin is quite low
in comparison, since the LD_{50} for the mouse is 280 mg/kg (Hofmann 1960).

Discovery of Psilocybin and Psilocin

The use of certain mushrooms containing psilocybin and psilocin by the
Aztec of Mexico dates back almost 2,000 years, and there is some evidence
that the Maya of Central America may have also used these fungi for their
hallucinogenic effects (Schultes and Hofmann 1973). The Aztec used the
mushrooms in various religious ceremonies and referred to them as *teonan-*

actl, or "flesh of the gods." Through the visions produced by the mushrooms, communication with the spirit world was believed to occur. Divination, prophecy, and curing rites likewise utilized the effects of the mushrooms. The use of these hallucinogens was despised by the Spanish conquerors of Mexico, and every effort was made to discourage their use and convert the natives of Mexico to Christianity. These efforts were largely successful, so that use of the mushrooms soon became restricted to certain isolated areas of Mexico. Aside from the early Spanish reports concerning the native use of these hallucinogenic fungi, very little was known about the practice or the mushrooms involved until the 1950s. Then, largely through the ethnobotanical investigations of Wasson and Heim (Heim and Wasson 1958; Heim 1963, 1967), the identity of some of the mushrooms utilized in Mexico for their hallucinogenic effects was established. The mushrooms were primarily species of *Psilocybe*, but species of *Conocybe* and *Stropharia* were also utilized for similar purposes.

Once the mushrooms had been identified, investigations began to characterize the active constituent. Albert Hofmann, of LSD fame, become involved in this study. Hofmann and his associates found that some species of these mushrooms could be grown in culture from spores. One species, *Ps. mexicana* Heim, was especially productive and was used for the isolation work. Laboratory animal-based bioassays for the mushroom's hallucinogenic effects proved to be difficult to develop, and thus the investigators themselves ingested the mushrooms, and fractions thereof, to determine their activity (Hofmann 1960; Schultes and Hofmann 1973). In 1958 the identity of the active constituents, psilocin and psilocybin, was reported (Hofmann et al. 1958a, b; 1959; Hofmann and Troxler 1959). Hofmann's group also prepared a large number of synthetic analogs of these unusual 4-substituted indole derivatives (Troxler, Seeman, and Hofmann 1959). Since the discovery of these compounds, numerous studies have shown that they are also present in many species of *Psilocybe*, *Stropharia*, *Conocybe*, *Copelandia*, and *Panaeolus* (Schultes and Hofmann 1973; Benedict et al. 1962; Heim, Hofmann, and Tscherter 1966; Benedict, Tyler, and Watling 1967; Ola'h 1968; Picker and Richard 1970; Ola'h 1970; Guzman and Ott 1976; Ott and Guzman 1976). Psychoactive species of these genera are not restricted to Mexico and have been reported from North and South America, Europe, Asia, Australia, and Africa.

Psilocybin is relatively resistant to degradation and has been detected in herbarium specimens over twenty-five years old. In contrast, psilocin is easily oxidized to inactive compounds and is thus usually present in much lower levels. Dried carpophores of *Ps. mexicana* usually contain 0.2 to 0.4 percent psilocybin. An effective dose of the dried mushroom would thus be between 2 and 4 g.

Baeocystin and Norbaeocystin

Baeocystin and norbaeocystin, the monomethyl and demethyl derivatives of psilocybin, have also been detected in certain *Psilocybe* species (Leung and Paul 1968; Repke and Leslie 1977). These compounds have always been found to occur with psilocybin, and this has led to speculation that phosphorylation may take place prior to N-methylation in the biosynthesis of psilocybin from tryptophan (Leung and Paul 1968; Agurell and Nilsson 1968a, b). Baeocystin and norbaeocystin are usually present in low levels relative to psilocybin, and although little is known about their pharmacologic activity, these constituents probably contribute little to the effects of the mushrooms containing them. *Ps. baeocysis* Singer and Smith, which contains these compounds, has been reported to cause an unusually toxic reaction in children, one case being fatal (McCawley, Brummett, and Dara 1962). However, the identity of the mushroom(s) involved in these cases was not established with certainty.

Pharmacologic Effects of Psilocybin and Psilocin

The effects of psilocybin have been studied extensively in man, and they are very similar to those of LSD (Hollister 1961; Hollister and Hartman 1962; Wolbach, Miner, and Isbel 1962). However, the duration of action is shorter and the potency of psilocybin is lower relative to that of LSD. Psilocybin appears to be hydrolyzed to psilocin *in vivo*, and the latter compound may actually be responsible for the CNS effects (Horita and Weber 1961a, b). Psilocybin is readily cleaved by alkaline phosphatase, which is found in many tissues, and when the conversion to psilocin is inhibited, the effects of psilocybin in mice are reduced (Horita and Weber 1962). In addition, psilocin is more lipophilic than psilocybin, which suggests that it would have greater CNS activity. The effects of psilocin are probably related to its structural similarity to serotonin, a neurotransmitter in the CNS.

When administered to rats, about 75 percent of a dose of psilocin is metabolized prior to excretion (Kalberer, Kreis, and Rutschmann 1962). Although little is known about the metabolism of psilocin, several studies have shown that cytochrome oxidase (Weber and Horita 1963) and ceruloplasmin (Levine 1967) can oxidize psilocin to a blue metabolite (perhaps an ortho-quinone). It is interesting to note that almost all mushrooms that contain psilocybin will develop a blue coloration when injured (Benedict, Tyler, and Watling 1967). This reaction is so distinct for *Psilocybe* species that many individuals using these mushrooms for recreational purposes consider it to be an infallible identifying feature of psilocybin-containing carpophores (Pollock 1975).

Recreational Use of Psilocybin-Containing Species

Psilocybin-containing mushrooms are becoming popular recreational drugs and can be found by the knowledgeable collector in many areas of the United States. The species most frequently used are *Stropharia cubensis* Earle (Gulf Coast states), *Panaeolus cyanescens* Berk and Br. (Florida), *Ps. pelliculosa* Singer and Smith, and *Ps. semilanceata* (Fr. ex Secr.) Kummer (Northwest and Canada) (Ott 1976b). Alleged psilocybin-containing mushrooms are also being sold on the illicit market. In the past the majority of these drugs contained only LSD or a combination of LSD and PCP (phencyclidine) (Brown and Malone 1976). This situation may change in the future now that home cultivation methods for several psilocybin-containing mushrooms have been published (Oss and Oeric 1976). Spores of these fungi are available through numerous advertisements in publications such as *High Times*.

Treatment

Adverse reactions from psilocybin-containing mushrooms are relatively rare. Since the duration of action is short, specific therapy is rarely needed. The patient can usually be talked through the hallucinogenic experience; but if a panic reaction develops, a tranquilizer, such as chlorpromazine, can be used to calm the patient and terminate the effects of psilocybin.

A potential problem that could be life threatening is the possible confusion of psilocybin-containing LBMs ("little brown mushrooms") with ama-toxin-containing LBMs, namely, *Galerina* and *Conocybe* species. The latter genus is particularly dangerous, since *Conocybe filaris* Fr. contains the ama-toxins (Brady et al. 1975) while *C. cyanopus* (Atk.) Kuhner and *C. smithii* Watling (Benedict et al. 1962) contain psilocybin.

Gymnopilus spectabilis

Another mushroom that has been reported to produce hallucinogenic ef-fects that are very similar to those of psilocybin is *Gymnopilus spectabilis* (Fr.) A. H. Smith (Walters 1965; Buck 1967; Sanford, 1972). In 1967 Buck re-ported that a collection of this mushroom, which had hallucinogenic activ-ity, was devoid of psilocybin. This report was accepted as evidence that the active constituent was distinct from psilocybin. However, recently psilocybin was detected in this mushroom as well as in eighteen other species of *Gym-nopilus* (Hatfield, Valdes, and Smith 1978). In this study it was also found that psilocybin was inconsistent in its occurrence in *Gymnopilus* species. A similar situation has been noted for the genus *Panaeolus* (Ola'h 1970).

Muscarine

Symptoms of Intoxication

As mentioned previously, muscarine is a potent parasympathomimetic that was first isolated from *Amanita muscaria*. This mushroom does not contain toxicologically significant levels of muscarine, but a number of mushrooms do, and produce the classical symptoms of muscarine intoxica-tion when ingested. The effects of muscarine reflect its potent cholinergic activity at postganglionic parasympathetic synapses. The onset of symptoms is usually within 15 to 30 minutes and rarely after 1 hour. The most com-monly reported symptoms are sweating, nausea, vomiting, and abdominal colic. Occasionally reported symptoms include blurred vision, salivation, rhinorrhea, and diarrhea. In severe cases asthma-like symptoms, bradycar-dia, hypotension, and shock develop. The mortality rate is approximately 5 percent, and when death does occur, it usually takes place within 8–9 hours (Waser 1961). Several cases of muscarine intoxication have been reported in the literature (Ford and Sherrick 1911; Clark and Smith 1913; Jelliffe 1937; Wilson 1947; Varese and Barbero 1967; Maretic 1967). The lethal dose of muscarine depends on the route of administration. In the mouse the LD_{50} of muscarine chloride is approximately 0.2 mg/kg when administered intravenously, but muscarine is much less toxic when given orally (Fraser 1957).

Structure of Muscarine

The potent pharmacologic effects of muscarine, which was the first substance known to produce stimulation of the parasympathetic nervous system, led to extensive research on its chemistry and pharmacology. Determination of the structure proved to be a difficult task and culminated after almost 150 years with the X-ray crystallographic analysis of muscarine iodide (Jellinek 1957; Kogl et al. 1957). The history, chemistry, and pharmacology of muscarine and related compounds have been reviewed by Wilkinson (1961) and Waser (1961).

Recently an investigation of the biosynthesis of muscarine by cultures of *Clitocybe rivulosa* Kummer was published (Mitta, Stadelmann, and Eugster 1977). Pyruvate was found to be incorporated into the C2 methyl group as well as C2 and C3 of the tetrahydrofuran ring, while glutamate carbons 2, 3, and 4 were incorporated into C4, C5, and the C5 methylene of muscarine. A biosynthetic relationship between muscarine and the isoxazole derivatives, ibotenic acid and muscimol, was also proposed. The latter compounds may also be derived from glutamic acid.

Mushrooms Containing Muscarine

Muscarine and its stereoisomers (epimuscarine, allomuscarine, and epiallomuscarine) have been found to occur in numerous genera of mushrooms (Catalfomo and Eugster 1970; Stadelmann, Muller, and Eugster 1976). However, in most cases the concentration of these compounds is too low (≤ 0.02 mg/g dry weight) to be of toxicological significance (Stadelmann, Muller, and Eugster, 1976). The only species that contain enough muscarine to cause intoxications are in the genera *Clitocybe* and *Inocybe*.

Several investigations dealing with the quantitative determination of muscarine in *Incoybe* species have been published (Eugster 1968; Catalfamo and Eugster 1970; Stadelmann, Muller and Eugster 1976; Eugster and Muller 1959; Eugster 1957; Malone et al. 1961; Brown et al. 1962). *I. napipes* Lange contains the highest level of muscarine, 0.71 percent (dry weight), according to a chromatographic assay (Brown et al., 1962). Catalfomo and Eugster (1970) determined the distribution and relative concentration of muscarine and its stereoisomers in several *Inocybe* species. In most cases muscarine was the predominant isomer present. However, in some species the amount of the epi-isomer ranged up to 58 percent. The epiallo- and allo-isomers were also found, but usually only in trace amounts. The toxicological significance of the occurrence of muscarine isomers is probably not great, since muscarine is much more potent in its pharmacologic effects than its isomers (Waser 1961). The genus *Inocybe* is composed of rather nondescript, generally brownish agaries, often with characteristic odors. Even though it is reasonably easy to determine whether a mushroom belongs to this genus, species designation is often difficult. The genus is widely distributed and usually is found in deep forests and other undisturbed areas. This habitat minimizes the chance of accidental ingestion by unattended children, and the brownish, fibrous appearance of most species in this genus has no appeal to the mycophagist. These factors are probably responsible for the relatively low incidence of *Inocybe* intoxications in the United States.

The genus *Clitocybe* is characterized by carpophores having white spores, fleshy central stipes, and broad adnate to decurrent gills. As with the *Inocybe*, the problems of identification and proper nomenclature are complex. Relatively few *Clitocybe* species have been assayed for muscarine content (Stadelmann, Muller, and Eugster 1976; Hughes, Genest, and Rice 1966; Genest, Hughes, and Rice 1968). Generally, these mushrooms contain less muscarine than is found in *Inocybe* species. For example, *C. dealbata* (Sow. ex Fr.) Kummer contains 0.15 percent muscarine (dry weight) (Hughes, Genest, and Rice 1966). As with *Inocybe* species, isomers of muscarine have also been found in *Clitocybe* (Stadelmann, Muller, and Eugster 1976).

Treatment

Unlike most other types of mushroom poisoning, muscarine intoxication has a specific antidote, atropine. Sufficient quantities should be given to produce dryness of the mouth. Larger quantities, sufficient to produce pupillary dilation, which have previously been advocated, are not necessary in most cases and may cause symptoms of atropine intoxication (delirium and hyperthermia) in small children.

Gyromitrin

Symptoms of Intoxication

One of the best-known and most sought-after groups of edible fungi are the true morels—ascomycetes of the genus *Morchella*. A closely related group of mushrooms are the so-called false morels. These fungi are also ascomycetes (*Gyromita*, *Helvella*, and *Verpa* species) and are called false morels because their carpophores are grossly similar to those of *Morchella* species. One of the most common false morels is *Gyromitra esculenta* Fr., and in spite of its species epithet this mushroom has caused numerous intoxications, some of which have been fatal. The onset of symptoms is usually 2 to 8 hours after ingestion and includes fatigue, headache, dizziness, and a feeling of fullness and/or abdominal pain. These symptoms are usually accompanied by nausea and vomiting, which may persist for several hours. In severe intoxications fever, convulsions, dyspnea, and symptoms of liver injury are also seen. Autopsies following fatal cases have shown fatty degeneration of the liver, kidneys, and myocardium. Franke, Freimuth, and List (1967) have reviewed over 500 cases of *Gyromitra* intoxication and found that about 15 percent were fatal. Death is usually due to severe liver damage.

An aspect of *G. esculenta* toxicity that has caused some confusion is that the mushroom appears to have variable toxicity. Most cases of *G. esculenta* intoxication have been reported from Eastern European countries, particularly Poland (Franke, Freimuth, and List, 1967). In contrast, *G. esculenta* is considered to be an edible mushroom in the western United States and is frequently collected and eaten, with no reports of adverse effects. In the Midwest and eastern United States, the situation appears to be intermediate between these two extremes. The mushroom is generally regarded as being edible by the public, but several poisoning cases have been reported from this region, with seven fatalities (Dearness 1924; Hendricks 1940; Cottingham 1955). It has been proposed that different chemical strains of this species exist, but this suggestion has not been verified by chemical or toxicological analysis.

Hydrazones of *Gyromitra esculenta*

In 1967 the principal toxic constituent of European G. *esculenta* was identified as acetaldehyde N-methyl-N-formylhydrazone (AMFH) by List and Luft (1967, 1968). This compound was given the common name gyromitrin (see figure 2.6). Several other related hydrazones were recently identified by Pyysalo (Pyysalo 1975, 1976; Pyysalo and Niskanen 1977). In all,

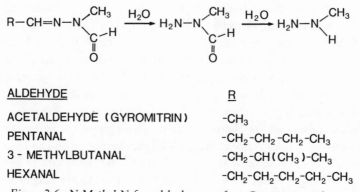

ALDEHYDE	R
ACETALDEHYDE (GYROMITRIN)	$-CH_3$
PENTANAL	$-CH_2-CH_2-CH_2-CH_3$
3 - METHYLBUTANAL	$-CH_2-CH(CH_3)-CH_3$
HEXANAL	$-CH_2-CH_2-CH_2-CH_2-CH_3$

Figure 2.6. N-*Methyl*-N-*formyl hydrazones from* Gyromitra esculenta.

nine N-methyl-N-formylhydrazones of simple aliphatic aldehydes have been detected in the mushroom. Early reports (List and Luft 1967, 1968) stated that carpophores collected in eastern Europe contained 1.2 to 1.6 g of gyromitrin (and analogs) per kg of fresh mushroom. However, this determination was made using a rather nonspecific potassium iodate assay for methylhydrazine. A more specific analytic method (gas chromatography using a capillary column) has since been developed, and using this assay the abundance of these compounds was found to be 57 mg/kg (fresh weight) (Pyysalo and Niskanen, 1977). Over 75 percent of the hydrazone fraction consisted of AMFH. This concentration correlated well with the toxicity of the mushrooms assayed.

All of the hydrazones found in G. *esculenta* thus far yield N-methyl-N-formylhydrazine (MFH) and N-methylhydrazine (MH) when subjected to hydrolysis (see figure 2.6). Nagel, Toth, and Kupper (1976) found that when AMFH is dissolved in a 37° buffer, pH 4 or lower, it is converted quantitatively in seconds to MFH. N-methyl-N-formylhydrazine is much more resistant to hydrolysis, since it was found that a 1×10^{-3}M solution decomposed to methylhydrazine with a half-life of 122 minutes at pH 2, 37°. Thus, orally ingested AMFH would be rapidly cleaved to MFH in the acid milieu of the stomach, and some MH would also be formed under these conditions. Once it has been absorbed, hydrolysis of AMFH appears to be much slower since Pyysalo and associates (Niskanen et al. 1976) recently reported detectable levels of this compound in the urine of a rabbit eleven days after the parenteral administration of a sublethal dose. The susceptibility of the gyromitrin analogs to hydrolysis is probably similar. In summary, the toxic hydrazones of *G. esculenta* are at least partially converted to MFH and MH when they are ingested orally, and the toxicity of these hydrolytic products must be considered when evaluating the oral toxicity of this mushroom.

Toxicity

Several studies of the acute toxicity of gyromitrin in various laboratory animals have been published (List and Luft 1968; Pyysalo and Niskanen 1977; Niskanen et al. 1976; Schmidlin-Meszaroa 1974). The lethal dose for rabbits is approximately 70 mg/kg; for rats, about 320 mg/kg; and for chickens, over 400 mg/kg. Several AMFH analogs have also been tested for toxicity and have been shown to have lower toxicity in rabbits than gyromitrin itself (Pyysalo 1975). The lethal dose in humans has been estimated to be between 10 and 50 mg/kg (Schmidlin-Meszaroa 1974).

The acute toxicity of MFH has not been studied as thoroughly. Toth and Erickson (1977) found that a dose of 300 mg/kg (SQ) produced death in 100 percent of the mice treated. Lethal doses of this compound did not produce convulsions, which is a common toxic effect of many hydrazines. In the same study a dose of 40 mg/kg (SQ) of methylhydrazine was found to produce 100 percent mortality in mice. Convulsive seizures were exhibited before death.

An unusual aspect of the acute toxicity of hydrazine derivatives is the "effect/no effect" dose–response relationship exhibited. This effect may be responsible for the fact that children are more frequently the victims of *Gyro-*

mitra intoxication than adults. Children are not more sensitive to the hydrazines, but they usually ingest more toxic material per kg of body weight and thus exceed the threshold dose.

Mechanism of Action and Treatment

It has been suggested that the convulsions produced by hydrazines may be due to inhibition of enzyme systems requiring pyridoxal phosphate (vitamin B_6) (Killam and Bain 1957). It is known that 1,1-dimethylhydrazine forms a hydrazone with pyridoxal phosphate *in vivo*, and this derivative has also been shown to be toxic (Cornish 1969; Dubnick, Leeson, and Scott 1960; Furst and Gustavson 1967). It is possible that MH, and perhaps MFH, acts in a similar way. Pyridoxine hydrochloride has been found to be an effective antidote for hydrazine (Kiklin et al. 1976) and methylhydrazine (Toth and Erickson 1977). However, the toxic effects produced by MFH are only slightly reduced by this vitamin (Toth and Erickson 1977). This observation, along with the absence of convulsions with MFH, indicates that this compound is not toxic because of its hydrolysis *in vivo* to MH.

Because of its effectiveness in treating hydrazine intoxications, the use of pyridoxine hydrochloride (25 mg/kg over 3 hours) has been suggested as an antidote for *Gyromitra* poisoning. The effectiveness of this therapy has not been tested in animals treated with lethal doses (oral and parenteral) of gyromitrin. However, considering the low cost and risk of such therapy, its use is recommended. All of the toxic effects may not be reversed by B_6, but the chances of convulsions would appear to be reduced.

It is interesting to note that pyridoxine hydrochloride has been shown to be an effective antidote in treating mice intoxicated by agaritine, β-N-[γ-L(+)-glutamyl]-4-hydroxymethylphenylhydrazine (Toth and Erickson 1977). This hydrazine derivative is found in *Agaricus bisporus* (Lange) Imbach, the commonly eaten commercial mushroom (Levenberg 1960a, b; Daniels, Kelly, and Hinman 1961). The mushrooms yield up to 0.04 percent (fresh weight) of this compound. In mice 200 mg/kg (SQ) produced death in 90 percent of the animals treated with agaritine, and convulsions were produced (Toth and Erickson 1977).

The hepatocellular damage that is occasionally produced by *G. esculenta* may be due, at least in part, to microsomal oxidation of MH in the liver.

Studies concerning the hepatotoxicity of isoniazid and iproniazid, two hydrazine derivatives used medicinally, have shown that monoalkyl and monoacyl hydrazines are oxidized by cytochrome P-450 to N-hydroxy intermediates that undergo dehydration to diazines (Mitchell et al. 1975; Nelson et al. 1976). These metabolites are thought to act as alkylating or acylating agents of hepatocyte macromolecules, an effect that could cause hepatic necrosis.

Detoxification of Gyromitra esculenta

The chronic effects of AMFH may also be important because "detoxified" G. esculenta is consumed in large quantities in Europe. The mushrooms are dried and/or cooked before ingestion. Because of the volatility of AMFH and related compounds, this processing removes most, but not all, of the hydrazine derivatives present (Pyysalo 1976; Pyysalo and Niskanen 1977; Schmidlin-Meszaroa 1974). Mushrooms that are air-dried at 15–20° for 5 days still contain over 12 percent of the original hydrazones present (about 45 mg/kg dry weight) (Pyysalo and Niskanen 1977). When they are dried for 14 days, the content drops to approximately 6 mg/kg dry weight. The most effective method of removing these compounds is to boil the carpophores. Boiling for 10 minutes reduces the hydrazone content to 0.53 percent of the original concentration (0.77 mg/kg undried) (Pyysalo and Niskanen 1977). Of course, boiling certainly would have an adverse effect on the distinct flavor and aroma that make the mushroom a desirable food item (Pyysalo 1976). Regardless of the method used to process G. esculenta, small amounts of the hydrazones remain in the carpophores, and thus the effects of chronic exposure to these compounds are of concern. In a 90-day "short-term" test, oral administration of AMFH was employed and the "no effect" values obtained were 0.05 mg/kg/day for chickens and 0.5 mg/kg/day for rabbits (Niskanen et al. 1976). When these results were applied to man, using a safety factor of 100, the maximum permissible amount of AMFH that could be safely ingested by a 70-kg subject would be approximately 0.035 mg/day. This amount would be found in 5 g of dried mushroom (dried 14 days) or 100 g of fresh mushroom that had been boiled 10 minutes. These values may not be overly conservative, since both MFH and MH have been recently shown to be tumorigenic (Toth 1972; Toth and Shimizu 1973;

Toth and Nagel 1977a). A derivative of agaritine, N-acetyl-4-hydroxy-methylphenylhydrazine, has also been shown to be carcinogenic (Toth and Nagel 1977b).

Coprine

Symptoms of Intoxication and Treatment

One of the most unusual types of mushroom intoxication is that produced by *Coprinus atramentarius* Fr. Ingestion of this mushroom with ethanol has been reported to cause a toxic reaction. Symptoms include profound flushing, a metallic taste, paresthesia in the extremities, palpitations, hyperventilation, hypotension, tachycardia, nausea, and vomiting. When the mushroom is not ingested with ethanol, these effects are not seen. Reynolds and Lowe (Reynolds and Lowe, 1965) have provided an excellent account of four cases of this type of poisoning. Mayer and associates reported one case in which the violent vomiting produced by the intoxication resulted in an esophageal rupture (Mayer, Herlocher, and Parisian 1971). The *Coprinus*–ethanol reaction has never been reported to be fatal, and no particular therapy is required in treating patients for this form of mushroom poisoning. Recovery is usually complete within several hours.

The common name of *C. atramentarius* is inky cap, which refers to the fact that the gills of this mushroom autolyse or deliquesce at maturity to form an inky fluid. This process is characteristic of most members of the genus *Coprinus* and is thought to be involved with spore release. *C. atramentarius* is very common in many areas of the United States, as are several other members of the genus, such as *C. comatus* (Mull. ex Fr.) S. F. Gray (shaggy mane) and *C. micaceus* (Bull. ex Fr.) Fries. The latter mushrooms do not produce the *Coprinus*–ethanol reaction.

Basis of the *Coprinus*–Ethanol Reaction

Reports of *C. atramentarius* intoxications prompted several studies concerning the mushroom's pharmacologic effects in animals (Barkman and Perman 1963; Genest, Coldwell, and Hughes 1968; Coldwell, Genest, and Hughes 1969). Of these investigations the most interesting was reported by

Coldwell, Genest, and Hughes (1969), who studied the effects of C. *atra-mentarius* on ethanol metabolism in mice. Administration of a mushroom extract followed by ethanol resulted in an increase in the blood level of acetaldehyde that was evident 15 minutes after ethanol was given and persisted throughout the 8-hour observation period. The fact that blood acetaldehyde levels were elevated gave an early indication of the basis of the *Coprinus*–ethanol reaction. Acetaldehyde, which is a metabolite of ethanol, is quite toxic, and blood levels above 5 μg/ml in humans usually produces effects similar to those reported for the *Coprinus*–ethanol reaction. Normally, the acetaldehyde produced by the hepatic oxidation of ethanol is rapidly metabolized to acetate, which is utilized in the citric acid cycle and various anabolic reactions. Even when large amounts of ethanol are ingested, blood levels of acetaldehyde normally remain below 1μg/ml. The liver is the principal site of acetaldehyde oxidation, and several aldehyde dehydrogenases have been isolated and studied from this organ (Deitrich and Hellerman 1963; Deitrich 1966; Deitrich and Kraemer 1968; Marjanen 1972; Tottmar, Pettersson, and Kiessling 1973; Crow et al. 1974; Horton and Barrett 1975). The enzyme primarily responsible for metabolism of acetaldehyde is an NAD-dependent dehydrogenase located within the mitochondria. Thus, it became apparent that the *Coprinus*–ethanol reaction was due to inhibition of the enzyme responsible for acetaldehyde oxidation. Since alcohol oxidation was not similarly inhibited, acetaldehyde accumulated, resulting in toxic levels of this metabolite.

Active Constituent of *Coprinus atramentarius*

A number of compounds are known to cause ethanol supersensitivity by inhibiting aldehyde dehydrogenase (Deitrich and Hellerman 1963). One such compound is disulfiram [*bis*-(diethylthiocarbamyl)-disulfide], which is used in the treatment of chronic alcoholism. When treated with disulfiram, the patient cannot adequately metabolize acetaldehyde when ethanol is ingested, and symptoms like those described for the *Coprinus*–ethanol reaction develop (Lundwall and Baekeland 1971). The possibility of such a reaction is often enough to deter an alcoholic from drinking.

The remarkable similarity between the *Coprinus*–ethanol and disulfiram–ethanol reactions gave rise to speculation that this compound might occur in the mushroom, and in 1956 Simandl and Franc (1956) reported the isolation of disulfiram. However, three independent reinvestigations failed to confirm this report (List and Reith 1960; Weir and Tyler 1960; Hatfield and Schaumberg 1975). Thus, disulfiram is definitely not responsible for the pharmacologic effects of C. *atramentarius*.

Recently the active constituent of the mushroom, coprine (see Figure 2.7), was isolated and identified by two groups working independently in the United States (Hatfield and Schaumberg 1975) and Sweden (Lindberg, Bergman, and Wickberg 1975, 1977; Lindberg 1977). The key to the success of

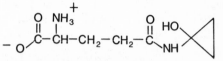

Figure 2.7. Coprine, the disulfiram-like constituent of Coprinus atramentarius.

these investigations was the development of bioassays for the *Coprinus*-ethanol reaction that could be used to guide fractionation of the mushroom. Hatfield and Schaumberg (1975) utilized the hyperaldehydemia produced in the reaction and were able to assess activity by assaying blood from treated mice for acetaldehyde using a gas chromatographic technique. A combination of anion-exchange, silica gel, and Sephadex chromatography was used to isolate the active compound, which was characterized primarily by its spectral properties.

Lindberg (1977) employed a different, but equally effective, bioassay for activity. He found that rats treated with C. *atramentarius*, or fractions thereof, and ethanol developed lachrymation and gradual swelling of the face, which reached a maximum about 18 hours after the ethanol was administered. This reaction could be graded as to intensity and was used as a guide in the isolation of coprine (Lindberg, Bergman, and Wickberg 1975, 1977; Lindberg 1977). An ethanol extract of the mushroom was fractionated primarily by a combination of cation- and anion-exchange chromatography. The mushroom yielded about 160 mg of coprine per kg (fresh weight). The yield was not affected by boiling or cooking the mushroom, but young carpophores were found to contain only about 80 mg/kg. Structural identification of coprine by these investigators was based on degradation and spectral studies as well as on synthesis of the compound. Wickberg and associates have also reported the synthesis of numerous analogs of coprine (Lindberg et al. 1977; Lindberg 1977).

This amino acid appears to be the first example of a naturally occurring compound containing a 1-aminocyclopropanol moiety. Several other amino acid derivatives containing a cyclopropane ring have been isolated from plant sources. One of the most interesting of these compounds is β-(methylenecyclopropyl)-alanine, or hypoglycin A (Hassall and Reyle 1955). When ingested, this amino acid is metabolized to 2-methylenecyclopropylacetic acid, which inhibits fatty acid oxidation, thus leading to depletion of liver glycogen (von Holt et al. 1964). Hypoglycin A has caused numerous cases of hypoglycemia in the West Indies, where large amounts of fruit (*Blighia sapida*) containing this compound are consumed.

Mechanism of Action and Treatment

The biochemical mechanism by which coprine acts has been studied, and it has been shown that the compound is an inhibitor of aldehyde dehydrogenase *in vivo* (Lindberg 1977; Tottmar and Lindberg 1977). This effect is obviously responsible for the *Coprinus*–ethanol reaction, as previous studies had indicated. However, coprine is inactive as an inhibitor of this enzyme *in vitro*, which indicates that an active inhibitor is formed from coprine *in vivo* (Hatfield and Schaumberg 1975; Lindberg 1977; Tottmar and Lindberg 1977). The active inhibitor appears to be 1-aminocyclopropanol, which can be formed from coprine by hydrolysis. Lindberg and associate (Lindberg 1977; Tottmar and Lindberg 1977) have shown that 1-aminocyclopropanol is active as an aldehyde dehydrogenase inhibitor both *in vivo* and *in vitro*. Some analogs of coprine have also been tested for activity, and only those compounds which can readily form a cyclopropaniminium ion or cyclopropanone are active both *in vivo* and *in vitro*. Derivatives that must be hydrolyzed to such compounds, such as coprine, are active only *in vivo*. Lindberg (1977) has proposed that cyclopropanone or a cyclopropaniminium ion, both of which could be formed from 1-aminocyclopropanol, reacts with one or more thiol groups in the active site of the enzyme, producing inactivation.

Coprine is a much more potent inhibitor of aldehyde dehydrogenase *in vivo* than disulfiram, but the duration of the effect appears to be about the same for both compounds (Lindberg 1977). In rats the peak of enzyme inhibition occurred about 6 hours after administration of coprine. At doses of 27 and 81 mg/kg, the activity did not return to normal even after 144 hours.

However, after 240 hours no inhibition was detected. Recovery is probably due to synthesis of new enzyme after excretion and/or metabolism of coprine and its hydrolytic product. The duration of inhibition in humans has not been published, but reactions to ethanol have been reported to occur as long as 48 hours after the mushroom was ingested.

Coprine appears to have a restricted distribution in the genus *Coprinus*. About 20 species have been screened for this compound, and thus far the only additional ones found to contain coprine are *C. quadrifidus* Peck, *C. variegatus* Peck, and *C. insignis* Peck (Hatfield, unpublished results). These species are generally less common than *C. atramentarius* in the United States.

Coprine may prove to be a useful drug in the treatment of alcoholism. Disulfiram does have a number of side effects, some of which are due to inhibition of dopamine-β-hydroxylase, an enzyme involved in the conversion of dopamine to norepinephrine (Goldstein et al. 1964). Coprine is not an inhibitor of this enzyme (Lindberg 1977). It is hoped that this mushroom metabolite may prove to be more specific in its action than disulfiram and thus cause fewer side effects in the treatment of alcoholism.

REFERENCES

Agurell, S. and J. Nilsson. 1968a. "A Biosynthetic Sequence from Tryptophan to Psilocybin." *Tetrahedron Lett.* 1968:1063–64.

—— 1968b. "Biosynthesis of Psilocybin." *Acta Chem. Scand.* 22:1210–18.

Allegro, J. 1970. *The Sacred Mushroom and the Cross.* New York-Doubleday.

Amalrac, F., M. Nicoloso, and J. Zalta. 1972. "A Comparative Study of 'Soluble' RNA Polymerase Activity of Zajdela Hepatoma Ascites Cells and Calf Thymus." *FEBS Lett.* 22:67–72.

Alleva, F. 1975. "Thioctic Acid and Mushroom Poisoning." *Science* 187:216.

Alleva, F., T. Balazs, A. Sager, and A. Done. 1975. "Failure of Thioctic Acid to Cure Mushroom Poisoned Mice and Dogs." Abstracts of Papers, Soc. Toxicology, 14th Annual Meeting, Williamsburg, Va.

Andary, C., F. Enjalbert, G. Privat, and B. Mandrou. 1977. "Assay for Amatoxins in *Amanita phalloides* with Direct Spectrometric Measurements of Chromatograms." *J. Chromatogr.* 132:525–32.

Barkman, R. and E. Perman. 1963. "Supersensitivity to Ethanol in Rabbits Treated with *Coprinus atramentarius*." *Acta Pharmacol. Toxicol.* 30:43–46.

Bartels, O. 1976. "Pilzvergiftungen." *Fortschr. Med.* 94:539–44.

Bartler, F. 1975. "Thioctic Acid and Mushroom Poisoning." *Science* 187:216.

Bastien, P. 1973. "Phalloide, es-tu enfin vaincue?" *Lyon Pharm.* 24:545–50.

Becker, C., T. Tong, U. Boerner, R. Roe, R. Scott, and M. MacQuarrie. 1976. "Diagnosis and Treatment of *Amanita phalloides*-type Mushroom Poisoning." *West J. Med.* 125:100–9.

Benedict, R., L. Brady, A. Smith, and V. Tyler, Jr. 1962. "Occurrence of Psilocybin and Psilocin in Certain *Conocybe* and *Psilocybe* Species." *Lloydia* 25:156–59.

Benedict, R., V. Tyler, Jr., and L. Brady. 1966. "Chemotaxonomic Significance of Isoxazole Derivatives in *Amanita* Species." *Lloydia* 29:333–41.

Benedict, R., V. Tyler, Jr., and R. Watling. 1967. "Blueing in *Conocybe*, *Psilocybe* and a *Stropharia* Species and the Detection of Psilocybin." *Lloydia* 30:150–57.

Benedict, R. 1972. "Mushroom Toxins Other Than *Amanita*." In S. Kadis, A. Ceigler, and S. Ajl, eds., *Microbial Toxins*, vol. 8: *Fungal Toxins*, pp. 281–320. New York: Academic Press.

Biggio, G., B. Brodie, E. Costa, and A. Guidotti. 1977. "Mechanism by Which Diazepam, Muscimol and Other Drugs Change the Content of cGMP in Cerebellar Cortex." *Proc. Natl. Acad. Sci. USA* 74:3592–96.

Block, S., R. Stephens, and W. Murrill. 1955. "The *Amanita* Toxins in Mushrooms." *J. Agric. Food Chem.* 3:584–87.

Bonetti, E., M. Derenzini, and L. Fiume. 1976. "Increased Penetration of Amanitine into Hepatocytes when Conjugated with Albumin." *Arch. Toxicol.* 35:69–73.

Bowden, K. and A. Drysdale. 1965. "A Novel Constituent of *Amanita muscaria*." *Tetrahedron Lett.* 1965:727–28.

Bowden, K., A. Drysdale, and G. Mogel. 1965. "Constituents of *Amanita muscaria*." *Nature (London)* 206:1359–60.

Brady, L. and V. Tyler, Jr. 1958. "A Chromatographic Examination of the Alkaloidal Fraction of *Amanita pantherina*." *J. Amer. Pharm. Assoc. Sci. Ed.* 48:417–19.

Brady, L., R. Benedict, V. Tyler, Jr., D. Stuntz, and M. Malone. 1975. "Identification of *Conocybe filaris* as a Toxic Basidiomycete." *Lloydia* 38:172–73.

Brehm, L., H. Hjeds, and P. Krogsgaard-Larsen. 1972. "The Structure of Muscimol, a GABA Analogue of Restricted Confirmation." *Acta Chem. Scand.* 26:1298–99.

Brodner, O. and T. Wieland. 1976. "Identification of the Amatoxin-Binding Subunit of RNA Polymerase by Affinity Labeling Experiments. Subunit B3—The True Amatoxin Receptor Protein of Multiple RNA Polymerase B." *Biochemistry* 15:3480–84.

Brown, J., M. Malone, D. Stuntz, and V. Tyler, Jr. 1962. "Paper Chromatographic Determination of Muscarine in *Inocybe* Species." *J. Pharm. Sci.* 51:853–56.

Brown, J. and M. Malone. 1976. "Status of Drug Quality in the Street-Drug Market—An Update." *Clin. Toxicol.* 9:145–68.

Buck, R. W. 1964. "Poisoning by Wild Mushrooms." *Clin. Med.* 71:1353–63.

—— 1967. "Psychedelic effects of *Pholiota spectabilis*." *N. Eng. J. Med.* 276:391–92.

Buku, A., G. Campadelli-Fiume, L. Fiume, and T. Wieland. 1971. "Inhibitory Effect of Naturally Occurring and Chemically Modified Amatoxins on RNA Polymerase of Rat Liver Nuclei." *FEBS Lett.* 14:42–44.

Buku, A., R. Altmann, and T. Wieland. 1974. "Components of the Green Death-cap Toadstool, Amanita phalloides. 46. The Nontoxic Sulfoxide Diastereoisomeric to O-Methyl-α-Amanitin." Justus Liebig's Ann. Chem. 1974: 1580–86.

Catalfomo, P. and C. Eugster. 1970. "Muscarine and Muscarine Isomers in Selected Inocybe Species." Helv. Chim. Acta 53:848–51.

Cavalli, P., R. Ragni, and G. Zuccaro. 1968. "L'Avvelenamento da funghi. Problemi terapeutici." Arch. Sci. Med. 125:690–51.

Cerletti, A. and A. Hofmann. 1963. "Mushrooms and Toadstools." Lancet 1:58–59.

Cessi, C. and L. Fiume. 1969. "Increased Toxicity of β-Amanitine when Bound to a Protein." Toxicon 6:309–10.

Chambon, P., G. Gissinger, J. Mandel, C. Kedinger, M. Gniazdowski, and M. Meilhac. 1970. "Purification and Properties of Calf Thymus DNA-Dependent RNA Polymerase A and B." Cold Spring Harbor Symp. Quant. Biol. 35:693–707.

Chilton, W. and J. Ott. 1976. "Toxic Metabolites of Amanita pantherina, A. cothurnata, A. muscaria and Other Amanita Species." Lloydia 39:150–57.

Ciocatto, E., U. Delfino, and M. Trompeo. 1970. "Trattamento nell'intossicazione da Amanita falloide e contributo clinico. Parte seconda." Minerva Anestesiol. 36:636–53.

Ciucci, N. and A. Chiri. 1974. "Considerazioni cliniche ed anatomapatologiche sui risulfate a distanze otlenati con l'acido tioctico nell'avvelenamento de Amanita phalloides." Minerva Anestesiol. 40:61–70.

Clark, E. and C. Smith. 1913. "Toxocological Studies of the Mushrooms Clitocybe illudens and Inocybe infida." Mycologia 5: 224–32.

Cochet-Meihlac, M. and P. Chambon. 1974. "Animal DNA-Dependent RNA Polymcrases. 2. Mechanism of the Inhibition of RNA Polymerase B by Amatoxins." Biochim. Biophys. Acta 353:160–83.

Coldwell, B., K. Genest, and D. Hughes. 1969. "Effect of Coprinus atramentarius on the Metabolism of Ethanol in Mice." J. Pharm. Pharmacol. 21:176–79.

Cornish, H. 1969. "The Role of Vitamin B_6 in the Toxicity of Hydrazines." Ann. N.Y. Acad. Sci. 166:136–45.

Cottingham, J. 1955. "Notes on Gyromitra esculenta Fr." Proc. Indiana Acad. Sci. 65:210–11.

Crow, K., T. Kitson, A. MacGibbon, and R. Batt. 1974. "Intracellular Localisation and Properties of Aldehyde Dehydrogenase from Sheep Liver." Biochem. Biophys. Acta 350:121–28.

Daniels, E., R. Kelly, and J. Hinman. 1961. "Agaritine: An Improved Isolation Procedure and Confirmation of Structure by Synthesis." J. Am. Chem. Soc. 83:3333–34.

Dearness, J. 1924. "Gyromitra Poisoning." Mycologia 16:199.

Deitrich, R. and L. Hellerman. 1963. "Diphosphopyridine Nucleotide-Linked Aldehyde Dehydrogenase." J. Biol. Chem. 238:1683–89.

Deitrich, R. 1966. "Tissue and Subcellular Distribution of Mammalian Aldehyde-Oxidizing Capacity." Biochem. Pharmacol. 15:1911–22.

Deitrich, R. and R. Kraemer. 1968. "Isolation and Characterization of Human Liver Aldehyde Adhydrogenase." *J. Biol. Chem.* 243:6402–8.

Delfino, U. 1970. "L'Intossicazione de *Amanita falloide.* Parte prima." *Minerva Anestesiol.* 36:629–35.

Dolara, P., E. Buiatti, and M. Geddes. 1971. "Hydrocortisone Protection of Phalloidin-Induced Rat Liver Lysosome Damage." *Pharmacol. Res. Commun.* 3:1–12.

Dubnick, B., G. Leeson, and C. Scott. 1960. "Effect of Forms of Vitamin B_6 on Acute Toxicity of Hydrazines." *Toxicol. Appl. Pharmacol.* 2:403–9.

Duffy, T. and P. Vergeer. 1977. *California Toxic Fungi.* San Francisco: Mycological Society of San Francisco.

Dyk, T., J. Piotriwska-Sowinska, and T. Buczkowska. 1969. "Proba oceny leczenia zatruc grzybami." *Pol. Tyg. Lek.* 24:1005–7.

Enna, S. and S. Snyder. 1976. "Gamma-Aminobutyric Acid (GABA) Receptor Binding in Mammalian Retina." *Brain Res.* 115:174–79.

Eugster, C. 1957. "Isolierung von Muscarin aus *Inocybe patouillardi* Bres." *Helv. Chim. Acta* 40:886–87.

Eugster, C. and G. Muller. 1959. "Notiz uber weitere Vorkommen von Muscarin." *Helv. Chim. Acta* 42:1189–1190.

Eugster, C., G. Muller, and R. Good. 1965. "Wirkstoffe aus *Amanita muscaria:* ibotensaure und muscazon." *Tetrahedron Lett.* 1965:1813–15.

Eugster, C. 1968. "Wirkstoffe aus dem Fliegenpilz." *Naturwissenschaften* 55:305–13.

Faulstich, H. and U. Fauser. 1973. "Investigations on the Question of Hemodialysis in *Amanita phalloides* Poisoning: Serum Levels and Excretion of Amanitin." *Dtsche. Med. Wochenschr.* 98:2258–59.

Faulstich, H. D. Georgopoulos, and M. Blocking. 1973. "Quantitative Chromatographic Analysis of Toxins in Single Mushrooms of *Amanita phalloides.*" *J. Chromatogr.* 79:257–65.

Faulstich, H., D. Georgopoulos, M. Blocking, and T. Wieland. 1974. "Analysis of the Toxins of Amanitin-Containing Mushrooms." *Z. Naturforsch.* 29C:86–88.

Faulstich, H. and M. Weckauf-Blocking. 1974. "Isolation and Toxicity of two Cytolytic Glycoproteins from *Amanita phalloides* Mushrooms." *Hoppe-Seyler's Z. Physiol. Chem.* 355:1489–94.

Faulstich, H., T. Wieland, A. Walli, and K. Birkmann. 1974. "Antamanides Protect Hepatocytes from Phalloidin Destruction." *Hoppe-Seyler's Z. Physiol. Chem.* 355:1162–63.

Faulstich, H., O. Brodner, S. Walch, and T. Wieland. 1975. "Uber die Inhaltsstoffe des grunen Knollenblatterpilzes. 49. Uber Phallisacin und Phallacin. Ein neue saure Phallotoxine und einege amide Phallacindins." *Justus Liebig's Ann. Chem.* 1975:2324–30.

Faulstich, H., H. Trischmann, and S. Zobeley. 1975. "A Radioimmunoassay for Amanitin." *FEBS Lett.* 56:312–15.

Faulstich, H. and M. Weckauf. 1975. "Cytolysis of Red Cells Mediated by Phallolysin, a Toxin Binding to N-Acetylglucosamine on the Cell Surface." *Hoppe-Seyler's Z. Physiol. Chem.* 356:1187–89.

Faulstich, H. and M. Cochet-Meilhac. 1976. "Amatoxins in Edible Mushrooms." *FEBS Lett.* 64:73–75.

Fauser, U. and H. Faulstich. 1973. "Beobachtungen zur Therapie die Knollenblatterpilzvergiftung: Besserung der Prognose durch Unterbrechung des enterhepatischen Kreislauf." *Dtsche. Med. Wochenschr.* 98:2259.

Finestone, A., R. Berman, B. Widmer. U. Laquer, and J. Markowitz. 1972. "Thioctic Acid Treatment of Acute Mushroom Poisoning." *Pa. Med.* 75:49–51.

Fiume, L. and R. Laschi. 1965. "Lesioni ultrastrutturali prodotte nelle cellule parenchimali spatiche dalla falloidina e dalla α-amanitinia." *Sperimentale* 115:288–97.

Fiume, L. and F. Stirpe. 1966. "Decreased RNA Content in Mouse Liver Nuclei After Intoxication with α-Amanitin." *Biochem. Biophys. Acta* 123:643–45.

Fiume, L., V. Marinozzi, and F. Nardi. 1969. "The Effects of Amanitin Poisoning on the Mouse Kidney." *Br. J. Exp. Pathol.* 50:270–76.

Fiume, L. 1972. "Pathogenesis of the Cellular Lesions Produced by α-Amanitin." In *The Biochemistry of Disease*, vol. 2, *The Pathology of Transcription and Translation*, pp. 105–22. New York: Marcel Dekker.

Fiume, L., M. Derenzini, V. Marinozzi, F. Petazzi and A. Testoni. 1973. "Pathogenesis of Gastrointestinal Symptomology During Poisoning with *Amanita phalloides*." *Experientia* 29: 1520–21.

Fiume, L., C. Busi, G. Campadelli-Fiume, and C. Franceschi. 1975. "Production of Antibodies to Amanitins as the Basis for Their Radioimmunoassay." *Experientia* 31:1233–34.

Fiume, L., S. Sperti, L. Montanaro, C. Basi, and D. Costantino. 1977. "Amanitins Do Not Bind to Serum Albumin." *Lancet* 1:1111.

Floersheim, G. 1971. "Antagonistic Effects of Phalloidin, α-Amanitin, and Extracts of *Amanita phalloides*." *Agents Actions* 2:142–49.

Floersheim, G., J. Schneeberger, and K. Bucher. 1971. "Curative Potencies of Penicillin in Experimental *Amanita phalloides* Poisoning." *Agents Actions* 2:138–41.

Floersheim, G. 1972a. "Antidotes to Experimental α-Amanitin Poisoning." *Nature (London), New Biol.* 236:115–17.

—— 1972b. "Curative Potencies Against α-Amanitin Poisoning by Cytochrome C." *Science* 177:808–9.

—— 1974. "Rifampicin and Cysteamine Protect Against the Mushroom Toxin Phalloidin." *Experientia* 30:1310–12.

Food and Drug Administration, 1977. *National Clearinghouse for Poison Control Centers Bulletin: Tabulation of 1975 Case Reports*. Bethesda, Md.

Ford, W. and J. Sherrick. 1911. "On the Properties of Several Species of the Polyporaceae and of a New Variety of *Clitocybe, Clitocybe dealbata-sudorifica* Peck." *J. Pharmacol. Exp. Ther.* 2:549–58.

Fournier, E. and J. Ripault. 1965. "Etude toxicologique et histologique de l'intoxication par *Amanite phalloide*." *Bull. Med. Legale* 8:268–82.

Franke, S., U. Freimuth and P. List. 1967. "Uber die Giftigkeit der Fruhjahrslorchel *Gyromitra (Helvella) esculenta* Fr." *Arch. Toxikol.* 22:293–332.

Fraser, P. 1957. "Pharmacological Actions of Pure Muscarine Chloride." *Br. J. Pharmacol.* 12:47–52.

Frimmer, M. and R. Kroker. 1973. "The Role of Extracellular K⁺-Concentration in Phalloidin Effects on Isolated Hepatocytes and Perfused Livers." *Naunyn-Schmiedeberg's Arch. Pharmacol. Exp. Pathol.* 278:285–92.

Frimmer, M. and E. Petzinger. 1975. "Phalloiden-Antagonisten. 4. Mitteilung: Thioctsaure, SH-Verbindungen, Rifampicin, Chloretika, Dexamethason, Ostradiol, Unspezifische Hemmstoffe und unwirksame Verbindungen." *Arzneim.-Forsch.* 25:1881–84.

Furst, A. and W. Gustavson. 1967. "A Comparison of Alkylhydrazines and Their B₆ Hydrazones as Convulsant Agents." *Proc. Soc. Exp. Biol. Med.* 124:172–75.

Gagneux, A., F Hafliger, R. Good, and C. Eugster. 1965a. "Synthesis of Panterine (Agarin)." *Tetrahedron Lett.* 1965:2077–79.

Gagneux, A., F. Hafliger, R. Meirer, and C. Eugster. 1965b. "Synthesis of Ibotenic Acid." *Tetrahedron Lett.* 1965:2081–84.

Gaultier, M., F. Fournier, P. Gervais, J. Frejaville, and F. Prieur. 1962. "Cinq cas d'intoxication par *l'Amanite phalloide*." *Bull. Mem. Soc. Med. Hop. Paris* 113:967–78.

Gazzard, B., M. Weston, I. Murray-Lyon, H. Flax, C. Record, B. Portmann, P. Langley, E. Dunlop, P. Mellon, W. Ward, and R. Williams. 1974. "Charcoal Haemoperfusion in the Treatment of Fulminant Hepatic Failure." *Lancet* 1:1301–7.

Genest, K., B. Coldwell and D. Hughes. 1968. "Potentiation of Ethanol by *Coprinus atramentarius* in Mice." *J. Pharm. Pharmacol.* 20:102–6.

Genest, K., D. Hughes and W. Rice. 1968. "Muscarine in *Clitocybe* Species." *J. Pharm. Sci.* 57:331–33.

Gerault, A. and L. Girre. 1975. "Recherches toxicologiques sur le genre *Lepiota* Fries." *C. R. Hebd. Séances Acad. Sci. Ser. D* 280:2841–44.

Goldstein, M., B. Anagnoste, E. Lauber, and M. McKereghan. 1964. "Inhibition of Dopamine-β-Hydroxylase by Disulfiram." *Life Sci.* 3:763–67.

Good, R., G. Muller and C. Eugster. 1965. "Isolierung und Charakterisierung von Pramuscimol und Muscazon aus *Amanita muscaria*." *Helv. Chim. Acta.* 48:927–30.

Goulon, M. 1967. "Oxygene hyperbare et fonction hepatique." *Ann. Anesthesiol. Fr.* 8:333–34.

Govindan, V., H. Faulstich, T. Wieland, B. Agostini, and W. Hasselbach. 1972. "*In vitro* Effect of Phalloidin on a Plasma Membrane Preparation from Rat Liver." *Naturwissenschaften* 59:521–22.

Govindan, V., G. Rohr, T. Wieland, and B. Agostini. 1973. "Binding of a Phallotoxin to Protein Filaments of Plasma Membrane of Liver Cell." *Hoppe-Seyler's Z. Physiol. Chem.* 354:1159–61.

Grossman, C. and B. Malbin. 1954. "Mushroom Poisoning: A Review of the Literature and Report of Two Cases Caused by a Previously Undescribed Species." *Ann. Intern. Med.* 40:249–59.

Guzman, G. and J. Ott. 1976. "Description and Chemical Analysis of a New

50   **G. M. Hatfield**

Species of Hallucinogenic *Psilocybe* from the Pacific Northwest." *Mycologia* 68:1261–67.
Hadjiolov, A., M. Dabeva, and V. Meckedonski. 1974. "The Action of α-Amanitin *in vivo* on the Synthesis and Maturation of Mouse Liver Ribonucleic Acids." *Biochem. J.* 138:321–33.
Harrison, D., C. Coggins, F. Welland, and S. Nelson. 1965. "Mushroom Poisoning in Five Patients." *Amer. J. Med.* 38:787–92.
Hassall, G. and K. Reyle. 1955. "Hypoglycin A and B, Two Biologically Active Polypeptides from *Blighia sapida*." *Biochem. J.* 60:334–39.
Hatfield, G. and L. Brady. 1975. "Toxins of Higher Fungi." *Lloydia* 38:36–55.
Hatfield, G. and J. Schaumberg. 1975. "Isolation and Structural Studies of Coprine, the Disulfiram-like Constituent of *Coprinus atramentarius*." *Lloydia* 38:489–96.
Hatfield, G., L. Valdes, and A. Smith. 1978. "The Occurrence of Psilocybin in *Gymnopilus* Species." *Lloydia* 41:140–44.
Hatfield, G., unpublished results.
Heim, R. and R. Wasson. 1958. *Les Champignons Hallucinogènes du Mexique*. Paris, Edit. Mus. Nat. Hist. Nat.
Heim, R. 1963. *Les Champignons Toxiques et Hallucinogènes*. Paris. N. Boubee.
Heim, R., A. Hofmann, and T. Tscherter. 1966. "Sur une intoxication collective à syndrome psilocybien causée en France par un *Copelandia*." *C. R. Hebd. Séances Acad. Sci. Ser. D* 262:519–24.
Heim, R. 1967. *Nouvelles Investigations sur les Champignons Hallucinogènes*. Paris, Edit. Mus. Nat. Hist. Nat.
Hendricks, H. 1940. "Poisoning by False Morel (*Gyromitra esculenta*)." *J. Am. Med. Assoc.* 116:1625.
Hjeds, H. and P. Krogsgaard-Larson. 1976. "Synthesis of Some 4-Aminoalkyl-5-Methyl-3-Isoxazoles Structurally Related to Muscimol and γ-Amino-Butyric Acid (GABA)." *Acta Chem. Scand. Ser. B* 30:567–73.
Hofmann, A., A. Frey, H. Ott, T. Petrzilka, and F. Troxler. 1958a. "Konstitutionsaufklorung und Synthese von Psilocybin." *Experientia* 14:397–99.
Hofmann, A., R. Heim, A. Brack, and H. Kobel. 1958b. "Psilocybin, ein psychotroper Wirkstoff aus dem mexikanischen Rauschpilz *Psilocybe mexicana* Heim." *Experientia* 14:107–9.
Hofmann, A. and F. Troxler. 1959. "Identifizierung von Psilocin." *Experientia* 15:101–81.
Hofmann, A., R. Brack, H. Kobel, A. Frey, H. Ott, T. Petrzilka, and F. Troxler. 1959. "Psilocybin and Psilocin." *Helv. Chim. Acta* 42:1557–81.
Hofmann, A. 1960. "Psychotomimetica. Chemische, pharmakologische und medizinische Aspekte." *Svensk. Kem. Tidskr.* 72:723–47.
Hollister, L. 1961. "Clinical, Biochemical and Psychologic Effects of Psilocybin." *Arch. Int. Parmacodyn.* 130:42–52.
Hollister, L. and A. Hartman. 1962. "Mescaline, Lysergic Acid Diethylamide and Psilocybin. Comparison of Clinical Syndromes, Effects on Color Perception and Biochemical Measures." *Comp. Psychiat.* 3:235–41.

Horgan, P., J. Ammirate and H. Thiers. 1976. "Occurrence of Amatoxins in *Amanita ocreata.*" *Lloydia* 39:368–71.

Horita, A. and L. Weber. 1961a. "Dephosphorylation of Psilocybin to Psilocin by Alkaline Phosphatase." *Proc. Soc. Exp. Biol. Med.* 106:32–34.

—— 1961b. "The Enzymatic Dephosphorylation and Oxidation of Psilocybin and Psilocin by Mammalian Tissue Homogenates." *Biochem. Pharmacol.* 7:47–54.

—— 1962. "Dephosphorylation of Psilocybin in the Intact Mouse." *Toxicol. Appl. Pharmacol.* 4:730–37.

Horton, A. and M. Barrett. 1975. "The Subcellular Localization of Aldehyde Dehydrogenase in Rat Liver." *Arch. Biochem. Biophys.* 167:426–36.

Hughes, D., K. Genest, and W. Rice 1966. "The Occurrence of Muscarine in *Clitocybe dealbata.*" *Lloydia* 29:328–32.

Jacob, S., F. Sajdel, W. Maecke, and H. Munro. 1970. "Soluble RNA Polymerases of Rat Nuclei: Properties, Template Specificity, and Amanitin Responses *in vitro* and *in vivo.*" *Cold Spring Harbor Symposium Quant. Biol.* 35:681–91.

Jacob, S., F. Sajdel and H. Munro. 1970. "Specific Action of α-Amanitin on Mammalian RNA Polymerase Protein." *Nature (London)* 225:60–62.

—— 1971. "Mammalian RNA Polymerases and Their Inhibition by Amanitin." *Adv. Enzyme Regul.* 9:169–81.

Jelliffe, S. 1937. "Some Notes on Poisoning by *Clitocybe dealbata* Sow. var. *sudorifica* Peck." *N.Y. State J. Med.* 37:1357–64.

Jellinek, F. 1957. "The Structure of Muscarine." *Acta Crystallogr.* 10:277–80.

Johnson, B., J. Preston, and J. Kimbrough. 1976. "Quantitation of Amatoxins in *Galerina autumnalis.*" *Mycologia* 68:1248–53.

Johnston, G., D. Curtis, W. deGroat and A. Duggan. 1968. "Central Actions of Ibotenic Acid and Muscimol." *Biochem. Pharmacol.* 17:2488–89.

Johnston, G. 1971. "Muscimol and the Uptake of γ-Amino-Butyric Acid by Rat Brain Slides." *Psychopharmacologia* 22:230–33.

Kalberer, F., W. Kreis, and J. Rutschmann. 1962. "The Fate of Psilocin in the Rat." *Biochem. Pharmacol.* 11:261–69.

Kedinger, C., M. Gniazdowski, J. Mandel, F. Gissinger, and P. Chambon. 1970. "α-Amanitin: A Specific Inhibitor of One of the Two DNA-Dependent RNA Polymerase Activities from Calf Thymus." *Biochem. Biophys. Res. Comm.* 38:165–71.

Killam, K. and J. Bain. 1957. "Convulsant Hydrazines I: *In Vitro* and *in Vivo* Inhibition of Vitamin B₆ Enzymes by Convulsant Hydrazines." *J. Pharmacol. Exp. Ther.* 119:255–62.

Kirklin, J., M. Watson, C. Bondoc, and J. Burke. 1976. "Treatment of Hydrazine-Induced Coma with Pyridoxine." *N. Engl. J. Med.* 294:938–39.

Kobert, R. 1891. "Ueber Pilzvergiftungen." *St. Petersburg Med. Wochenschr.* 51:463, 52:471.

Kogl, F., C. Salemink, H. Schouten, and F. Jellinek. 1957. "Uber Muscarin III." *Rec. Trav. Chim. Pays-Bas* 76:109–27.

Kostansek, E., W. Lipscomb, R. Yocum, and W. Thiessen. 1977. "The Crystal

Structure of the Mushroom Toxin β-Amanitin." *J. Am. Chem. Soc.* 99: 1273–74.

Krogsgaard-Larson, P. and G. Johnston. 1975. "Inhibition of GABA Uptake in Rat Brain Slices by Nipecotic Acid, Various Isoxazoles and Related Compounds." *J. Neurochem.* 25:797–802.

Krogsgaard-Larson, P., G. Johnston, D. Curtis, C. Game, and R. McCulloch. 1975. "Structure and Biological Activity of a Series of Conformationally Restricted Analogues of GABA." *J. Neurochem.* 25:803–9.

Krogsgaard-Larson, P. and S. Christensen. 1976. "Structural Analogues of GABA. A New Convenient Synthesis of Muscimol." *Acta Chem. Scand. Ser. B* 30:281–82.

Kubicka, J. 1963. "New Possibilities in the Treatment of Poisoning by the Deadly Amanita—*Amanita Phalloides.*" *Mykol. Mitteil.* 7:92–94.

Kubicka, J. and A. Adler. 1968. "Uber eine neuere Behandlungs methode der Vergiftung durch den Knollenblatterpilz." *Praxis* 57:1304–6.

Lengsfeld, A., I. Low, T. Wieland, and P. Dancker. 1974. "Interaction of Phalloidin with Actin." *Proc. Nat. Acad. Sci. USA* 71:2803–7.

Leung, A. and A. Paul. 1968. "Baeocystin and Norbaeocystin: New Analogues of Psilocybin from *Psilocybe baeocystis.*" *J. Pharm. Sci.* 57:1667–71.

Levenberg, B. 1960a. "Isolation and Enzymatic Reactions of Agaritine, a New Amino Acid from Agaricaceae." *Fed. Proc.* 19:6.

—— 1960b. "Structure and Enzymatic Cleavage of Agaritine, a New Phenylhydrazine of L-Glutamic Acid Isolated from Agaricaceae." *J. Am. Chem. Soc.* 83:503–4.

Levine, W. 1967. "Formation of Blue Oxidation Product from Psilocybin." *Nature (London)* 215:1292–93.

Lewis, B. 1955. "Atropine in Mushrooms." *S. African Med. J.* 29:262–63.

Limousin, H. 1959. "Le traitement de l'intoxication par les champignons vénéneux." *Sem. Hop. Paris* 35:827–29.

Lindberg, P., R. Bergman, and B. Wickberg. 1975. "Isolation and Structure of Coprine, a Novel Physiologically Active Cyclopropane Derivative from *Coprinus atramentarius*, and its Synthesis via 1-Aminocyclopropanol." *J. Chem. Soc. Chem. Commun.* 1975:946–47.

Lindberg, P. 1977. "Coprine, a Cyclopropanone-Related Disulphiram-like Constituent of *Coprinus atramentarius.*" Ph. D. dissertation, Lund Institute of Technology, Lund, Sweden.

Lindberg, P., R. Bergman, and B. Wickberg. 1977. "Isolation and Structure of Coprine, the *in Vivo* Aldehyde Dehydrogenase Inhibitor in *Coprinus atramentarius*; Synthesis of Coprine and Related Cyclopropane Derivatives." *J. Chem. Soc. Perkin Trans. I* 1977:684–91.

Lindell, T., F. Weinberg, P. Morris, R. Roeder, and W. Rutter. 1970. "Specific Inhibition of Nuclear RNA Polymerase II by α-Amanitin." *Science* 170:447–49.

List, P. and H. Reith. 1960. "Der Faltentinling, *Coprinus atramentarius* Bull., und seine dem tetraethylthiuramdisulfid ahnliche Wirkung." *Arzneim.-Forsch.* 10:34–40.

List, P. and P. Luft. 1967. "Gyromitrin, das Gift der Fruhjahrslorchel, *Gyromitra (Helvella) esculenta* Fr." *Tetrahedron Lett.* 1967:1893–94.

—— 1968. "Gyromitrin, das Gift der Fruhjahrslorchel." *Archiv. Pharmazie.* 301:294–305.

Loev, B., J. Wilson, and M. Goodman. 1970. "Synthesis of Compounds Related to Muscimol." *J. Med. Chem.* 13:738–41.

Lundwall, L. and F. Baekeland. 1971. "Disulfiram Treatment of Alcoholism." *J. Nerv. Ment. Dis.* 153:381–94.

Lutz, F., H. Glossman, and M. Frimmer. 1972. "Binding of ^{3}H-Desmethylphalloidin to Isolated Plasma Membranes from Rat Liver." *Naunyn-Schmiedeberg's Arch. Pharmacol.* 273:341–51.

Lykkeberg, J. and P. Krogsgaard-Larson. 1976. "Structural Analogues of GABA. Synthesis of 5-Aminoethyl-3-Isothiazolol (Thiomuscimol)." *Acta Chem. Scand. Ser. B.* 30:781–85.

Malone, M., R. Robichaud, V. Tyler, Jr., and L. Brady. 1961. "A Bioassay for Muscarine Activity and its Detection in Certain *Inocybe*." *Lloydia* 24:204–10.

—— 1962. "Relative Muscarinic Potency of Thirty *Inocybe* Species." *Lloydia* 25:231–37.

Maretic, Z. "Poisoning by the Mushroom *Clitocybe olearia* Maire." *Toxicon* 4:263–67.

Marjanen, L. 1972. "Intracellular Localization of Aldehyde Dehydrogenase in Rat Liver." *Biochem. J.* 127:633–39.

Mayer, J., J. Herlocher, and J. Parisian. 1971. "Esophageal Rupture After Mushroom Alcohol Ingestion." *N. Engl. J. Med.* 285:1323.

McCawley, E., R. Brummett, and G. Dana. 1962. "Convulsions from *Psilocybe* Mushroom Poisoning." *Proc. West. Pharmacol. Soc.* 5:27–33.

Meihlac, M., C. Kedinger, P. Chambon, H. Faulstich, M. Govindan, and T. Wieland. 1970. "Amanitin Binding to Calf Thymus RNA Polymerase B." *FEBS Lett.* 9:258–60.

Mitchel, D. 1976. "Mushroom Poisoning—Colorado Experience 1972–75." *Rocky Mountain Med. J.* 73:328–31.

Mitchell, J., U. Thorgeirsson, M. Black, J. Timbrell, W. Snodgrass, W. Potter, D. Jollow, and H. Keiser, 1975. "Increased Incidence of Isoniazid Hepatitis in Rapid Acetylators: Possible Relationship to Hydrazine Metabolites." *Clin. Pharmacol. Ther.* 18:70–79.

Mitta, K., R. Stadelmann, and C. Eugster. 1977. "Studies on the Biosynthesis of Muscarine in Mycelial Cultures of *Clitocybe rivulosa*." *Helv. Chim. Acta* 60:1747–53.

Moroni, F., R. Fantozzi, E. Masini, and P. Mannaioni. 1976. "A Trend in the Therapy of *Amanita phalloides* Poisoning." *Arch. Toxicol.* 36:111–15.

Muller, G. and C. Eugster. 1965. "Muscimol, ein pharmackodynamish wirksamer Stoff aus *Amanita muscaria*." *Helv. Chim. Acta* 48:910–26.

Myler, R., J. Lee, and J. Hooper, Jr. 1964. "Renal Tubular Necrosis Caused by Mushroom Poisoning." *Arch. Int. Med.* 114:196–204.

Nagel, D., B. Toth, and R. Kupper. 1976. "Formation of Methylhydrazine from Acetaldehyde N-Methyl-N-Formylhydrazone of *Gyromitra esculenta.*" *Proc. Am. Assoc. Cancer Res.* 17:9.

Nelson, S., J. Mitchell, J. Timbrell, W. Snodgrass, and G. Corcoran. 1976. "Isoniazid and Iproniazid: Activation of Metabolites to Toxic Intermediates in Man and Rat." *Science* 193:901–3.

Niskanen, A., H. Pyysalo, E. Rimaila-Parnanen, and P. Hartikka. 1976. "Short-Term Peroral Toxicity of Ethylene Gyromitrin in Rabbits and Chickens." *Food Cosmet. Toxicol.* 14:409–15.

Odenthal, K., R. Seeger, and G. Vogel. 1975. "Toxic Effects of Phallolysin from *Amanita Phalloides.*" *Naunyn-Schmiedeberg's Arch. Pharmacol.* 290:133–43.

Ola'h, G. 1968. "Etude chimotoxinomique sur les *Panaeolus*. Recherches sur la présence des corps indoliques psychotropes dans ces champignons." *C. R. Hebd. Séances Acad. Sci. Ser. D* 267:1369–72.

—— 1970. "Le Genre *Panaeolus:* Essai toxonomic et physiologique." *Rev. Mycologie, Memoire nos-serie,* no. 10.

Oss, O. and O. Oeric. 1976. *Psilocybin Magic Mushroom Growers Guide.* Berkeley: And/Or Press.

Ott, J., P. Wheaton, and W. Chilton. 1975. "Fate of Muscimol in the Mouse." *Physiol. Chem. Phys.* 7:381–84.

Ott, J. 1976a. "Psycho-mycological Studies of *Amanita*—From Ancient Sacrament to Modern Phobia." *J. Psyched. Drugs* 8:27–35.

—— 1976b. *Hallucinogenic Plants of North America.* Berkeley: Wingbow Press.

Ott, J. and G. Guzman. 1976. "Detection of Psilocybin in Species of *Psilocybe, Panaeolus* and *Psathyrella.*" *Lloydia* 39:258–60.

Paaso, B. and D. Harrison. 1975. "A New Look at an Old Problem: Mushroom Poisoning." *Amer. J. Med.* 58:505–9.

Palyza, V. and V. Kulhanek. 1970. "Uber die Chromatographische Analysis von Toxinen aus *Amanita phalloides.*" *J. Chromatogr.* 53:545–58.

Palyza, V. 1974. "Schnelle Identifizierung von Amanitinen in Pilzgeweben." *Arch. Toxicol.* 32:109–14.

Pfefferkorn, W. and G. Kirsten. 1967. "Beitrag zum Pantherina-Syndrom in Kindesalter—Pantherpilzvergiftung." *Kinderarztliche Praxis* 35:355–64.

Picker, J. and R. Richards. 1970. "The Occurrence of the Psychotomimetic Agent Psilocybin in an Australian Agaric, *Psilocybe subaeruginosa.*" *Aust. J. Chem.* 23:853–55.

Pollock, S. 1975. "The Psilocybin Mushroom Pandemic." *J. Psyched. Drugs* 7:73–84.

Preston, J., H. Stark, and J. Kimbrough. 1975. "Quantitation of Amanitins in *Amanita verna* with Calf Thymus RNA Polymerase B." *Lloydia* 38:153–61.

Pyysalo, H. 1975. "Some New Toxic Compounds in False Morels, *Gyromitra esculenta.*" *Naturwissenschaften* 62:395.

—— 1976. "Identification of Volatile Compounds in Seven Edible Fresh Mushrooms." *Acta Chem. Scand. Ser. B* 30:235–44.

Pyysalo, H. and A. Niskanen. 1977. "On the Occurrence of N-Methyl-N-Formyl-Hydrazones in Fresh and Processed False Morel, *Gyromitra esculenta.*" *J. Agr. Food Chem.* 25:644–47.

Raaen, H. 1969. "Improved Thin-Layer Chromatographic Separation of Amanita Toxins on Silica Gel G Chromatographic Plates." *J. Chromatogr.* 38:403–7.

Repke, D. and D. Leslie. 1977. "Baeocystin in *Psilocybe semilanceata.*" *J. Pharm. Sci.* 66:113–14.

Reynolds, W. and F. Lowe. 1965. "Mushrooms and a Toxic Reaction to Alcohol. Report of Four Cases." *N. Engl. J. Med.* 272:630–31.

Roberts, E. 1974. "γ-Aminobutyric Acid and Nervous System Function—A Perspective." *Biochem. Pharmacol.* 23:2637–49.

Salemink, C., J. ten Broeck, P. Schuller, and E. Veen. 1963. "Uber die basischen Inhaltsstoffe des Fliegenpilzes XII. Mitteilung: Uber die Anwesenheit von *l*-hyoscyamine." *Planta Med.* 11:139–41.

Sanford, J. 1972. "Japan's 'Laughing Mushrooms.' " *Econ. Bot.* 26:174–81.

Schmidlin-Meszaroa, J. 1974. "Gyromitrin in Trockenkorcheln (*Gyromitra esculenta*)." *Mitt. Geb. Lebensm. Hyg.* 65:453–65.

Schultes, R. and A. Hofmann. 1973. *The Botany and Chemistry of Hallucinogens.* Springfield, Ill.: Charles C Thomas.

Seeger, R. 1975a. "Demonstration and Isolation of Phallolysin, a Hemolytic Toxin from *Amanita phalloides.*" *Naunyn-Schmiedeberg's Arch. Pharmacol.* 287:277–87.

—— 1975b. "Some Physico-chemical Properties of Phallolysin Obtained from *Amanita phalloides.*" *Naunyn Schmiedeberg's Arch. Pharmacol.* 288:155–62.

Seeger, R., H. Kraus, and R. Wiedmann. 1975. "Zur Vorkommen von Hamolysin in Pilzen der Gattung *Amanita.*" *Arch. Toxikol.* 30:215–26.

Seeger, R. and O. Bartels. 1976. "Elimination of Toxic Peptides from *Amanita phalloides* by Charcoal Perfusion *in Vitro.*" *Dtsche. Med. Wochenschr.* 101:1456–58.

Seifart, K. and C. Sekeris. 1969. "α-Amanitin, a Specific Inhibitor of Transcription by Mammalian RNA-Polymerase." *Z. Naturforsch.* 24B:1538–44.

Seifart, K., B. Benecke, and P. Juhasz. 1972. "Multiple RNA Polymerase Species from Rat Liver Tissue: Possible Existence of a Cytoplasmic Enzyme." *Arch. Biochem. Biophys.* 151:519–32.

Siess, E., O. Wieland, and F. Miller. 1970. "Elektronenmikroskopische Untersuchunge zur Phalloidintoleranz neugeborenen Ratten, Mause and Kaninchen." *Virchows Archiv. B. Cell Pathol.* 6:151–65.

Simandl, J. and J. Franc. 1956. "Isolace tetraethylthiuramdisulfidn z hniku inkoustoveho (*Coprinus atramentarius*)." *Chem. Listy* 50:1862–63.

Sperti, S. L. Montanaro, L. Fiume, and A. Mattioli. 1973. "Dissociation Constants for Complexes Between RNA Polymerase II and Amanitins." *Experientia* 29:33–34.

Stadelmann, R., E. Muller, and C. Eugster. 1976. "Investigations on the Distribution of the Stereoisomeric Muscarines Within the Order of Agaricles." *Helv. Chim. Acta* 59:2432–46.

Stark, H., J. Kimbrough, and J. Preston. 1973. "Toxicological and Cultural Studies in the Genus *Amanita*." *ASB Bull*. 20:84–89.

Stiegers, C. and O. Strubelt. 1973. "Failure of Cytochrome C to Cure Experimental *Amanita phalloides* Poisoning." *Ger. Med*. 3:103–4.

Stirpe, F. and L. Fiume. 1967. "Studies on the Pathogenesis of Liver Necrosis by α-Amanitin." *Biochem. J*. 105:779–82.

Sullivan, G., L. Brady, and V. Tyler, Jr. 1965. "Identification of α- and β-Amanitin by Thin-Layer Chromatography." *J. Pharm. Sci*. 54:921–22.

Takemoto, T., T. Nakajima, and R. Sakuma. 1964. "Isolation of a Flycidal Constituent 'Ibotenic Acid' from *Amanita muscaria* and A. *pantherina*." *Yakugaku Zasshi* 84:1233–34.

Takemoto, T., T. Nakajima, and T. Yokoke. 1964. "Structure of Ibotenic acid." *Yakugaku Zasshi* 84:1232–33.

Talbot, G. and L. Vining. 1963. "Pigments and Other Extractives from Carpophores of *Amanita muscaria*." *Can. J. Botany* 41:638–47.

Tanghe, L. and D. Simons. 1973. "*Amanita phalloides* in the Eastern United States." *Mycologia* 65:99–108.

Theobald, W., O. Buch, H. Kinz, P. Krupp, E. Stenger, and H. Heimann. 1968. "Pharmakologische und experimental psychologische Untersuchungen mit 2 Inhaltsstoffen des Fliegenpilzes (*Amanita muscaria*)." *Arzneim-Forsch*. 18:311–15.

Toth, B. 1972. "Hydrazine, Methylhydrazine and Methylhydrazine Sulfate Carcinogenesis in Swiss Mice." *Int. J. Cancer* 9:109–18.

Toth, B. and H. Shimizu. 1973. "Methylhydrazine Tumorigenesis in Syrian Golden Hamsters and the Morphology of Malignant Histiocytomas." *Cancer Res*. 33:2744–53.

Toth, B. and J. Erickson. 1977. "Reversal of the Toxicity of Hydrazine Analogues by Pyridoxine Hydrochloride." *Toxicology* 7:31–36.

Toth, B. and D. Nagel. 1977a. "Cancer Induction with N-Methyl-N-Formylhydrazine, an Ingredient of the False Morel." *Second Int. Mycol. Congress, Abstracts*.

—— 1977b. "Tumorigenicity of the N-Acetyl Derivative of 4-Hydroxymethylphenylhydrazine, an Ingredient of *Agaricus bisporus*." *Proc. Am. Assoc. Cancer Res*. 18:15.

Tottmar, S., H. Pettersson, and K. Kiessling. 1973. "The Subcellular Distribution and Properties of Aldehyde Dehydrogenase in Rat Liver." *Biochem. J*. 135:577–86.

Tottmar, S. and P. Lindberg. 1977. "Effects on Rat Liver Acetaldehyde Dehydrogenase *in Vitro* and *in Vivo* by Coprine, the Disulfiram-like Constituent of *Coprinus atromentarius*." *Acta Pharmacol. Toxicol*. 40:476–81.

Troxler, F., F. Seemann, and A. Hofmann. 1959. "Abuandlungsprodukte von Psilocybin and Psilocin." *Helv. Chim. Acta* 42:2073–2103.

Truhaut, R., M. Thevenin, J. Warnet, and J. Claude. 1973. "Etude des relations possibles entre la steatose hepatique par l'alpha-amanitine et le metabolisme des lipoproteines chez le rat." *Ann. Biol. Clin*. 31:111–13.

Tuchweber, B., K. Kovacs, J. Khandekar, and B. Garg. 1972. "Effect of Phenobarbital and Steroids on Phalloidin Toxicity in Rats." *Toxicon* 10:357–61.

Tyler, V., Jr., 1958. "Pilzatropine, the Ambiguous Alkaloid." *Amer. J. Pharm.* 130:264–69.

—— 1963. "Poisonous Mushrooms." *Progr. Chem. Toxicol.* 1:339–84.

Tyler, V., Jr., L. Brady, R. Benedict, J. Khanna, and M. Malone. 1963. "Chromatographic and Pharmacologic Evaluation of Some Toxic *Galerina* Species." *Lloydia* 26:154–57.

Tyler, V., Jr. and A. Smith. 1963. "Chromatographic Detection of *Amanita* Toxins in *Galerina venenata*." *Mycologia* 55:358–59.

Tyler, V., Jr. and D. Groger. 1964. "Investigation of the Alkaloids of *Amanita* Species. I. *Amanita muscaria*." *Planta Med.* 12:334–39.

Tyler, V., Jr., R. Benedict, L. Brady, and J. Robbers. 1966. "Occurrence of *Amanita* Toxins in American Collections of Deadly Amanitas." *J. Pharm. Sci.* 55:590–93.

Varese, L. and S. Barbero. 1967. "Avvelenamento collettive de *Inocybe fastigiata*." *Minerva Pediatr.* 19:337–39.

von Holt, C., J. Chang, M. von Holt, and H. Bohm. 1964. "Metabolism and Metabolic Effects of Hypoglycin." *Biochim. Biophys. Acta* 90:611–13.

Walters, M. 1965. "*Pholiota spectabilis*, a Hallucinogenic Fungus." *Mycologia* 57:837–38.

Waser, P. 1961. "Chemistry and Pharmacology of Muscarine, Muscarone and Some Related Compounds." *Pharmacol. Rev.* 13:465–515.

—— 1967. "The Pharmacology of *Amanita muscaria*." In D. Efron, B. Holmstedt and N. S. Kline, eds., *Ethnopharmacologic Search for Psychoactive Drugs*, Public Health Service Publication no. 1645, pp. 419–38. Washington, D.C.: U.S. Government Printing Office.

Wasson, V. and R. Wasson. 1957. *Mushrooms, Russia and History*. New York: Pantheon Books.

Wasson, R. 1967. "Fly Agaric and Man." In D. Efron, B. Holmstedt and N. S. Kline, eds., *Ethnopharmacologic Search for Psychoactive Drugs*, Public Health Service Publication no. 1645, pp. 405–14. Washington, D.C.: U.S. Government Printing Office.

—— 1968. *Soma: Divine Mushroom of Immortality*. New York: Harcourt, Brace and World.

Weber, L. and A. Horita. 1963. "Oxidation of 4- and 5-Hydroxyindole Derivatives by Mammalian Cytochrome Oxidase." *Life Sci.* 2:44–49.

Weir, J. and V. Tyler, Jr. 1960. "An Investigation of *Coprinus atramentarius* for the Presence of Disulfiram." *J. Am. Pharm. Assoc. Sci. Ed.* 49:426–29.

Wieland, T. and O. Wieland. 1959. "Chemistry and Toxicology of the Toxins of *Amanita phalloides*." *Pharmacol. Rev.* 11:87–107.

Wieland, T., H. Schiefer, and U. Gebert. 1966. "Giftstoffe von *Amanita verna*." *Naturwissenschaften.* 53:39–40.

Wieland, T. 1967. "The Toxic Peptides of *Amanita phalloides*." *Fortsch. Chem. Org. Naturst.* 25:214–50.

—— 1968. "Poisonous Principles of Mushrooms of the Genus *Amanita*." *Science* 159:946–52.

58   G. M. Hatfield

Wieland, T., G. Luben, H. Ottenheym, J. Faesel, J. DeVries, A. Prox, and J. Schmid. 1968. "The Discovery, Isolation, Elucidation of Structure and Synthesis of Antamanide." *Angew. Chem. Int. Ed.* 7:204–8.

Wieland, T. 1972. "Isolierung und chemische Bearbeitung der Inhaltsstoffe des grunen Knollenblatterpilzes (*Amanita phalloides*)." *Arzneim.-Forsch.* 22:142–46.

Wieland, T., H. Faulstich, W. Jahn, M. Govindan, H. Puchinger, Z. Kopitar, H. Schmaus, and A. Schmitz. 1972. "Uber Antamanid. 14. Zur Wirkungsweise des Antamanides." *Hoppe-Seyler's Z. Physiol Chem.* 353:1337–45.

Wieland, T. and O. Wieland. 1972. "The Toxic Peptides of *Amanita* Species." In S. Kadis, A. Ciegler and S. Ajl, eds., *Microbial Toxins*, vol. 8. *Fungal Toxins*, pp. 249–80. New York: Academic Press.

Wieland, T. 1977. "Modification of Actins by Phallotoxins." *Naturwissenschaften* 64:303–9.

Wilkinson, S. 1961. "The History and Chemistry of Muscarine." *Quart. Rev. Chem. Soc.* 15:153–71.

Wilson, P. 1947. "Poisoning by *Inocybe fastigiata*." *Br. Med. J.* 1947:297.

Wolbach, A., E. Miner, and H. Isbel. 1962. "Comparison of Psilocin with Psilocybin, Mescaline and LSD-25." *Psychopharmacol.* 3:219–23.

Yocum, R. and D. Simons. 1977. "Amatoxins and Phallotoxins in *Amanita* Species of the Northeastern United States." *Lloydia* 40:178–90.

Zulik, R., F. Bako, and J. Badavari. 1972. "Death-Cap Poisoning." *Lancet* 2:228.

3 Toxins and Teratogens of the Solanaceae and Liliaceae

Richard F. Keeler

Many potentially poisonous plants from the Solanaceae and Liliaceae are commonly ingested as foods by domestic livestock and humans. Some are sources of drugs or drug precursors. This chapter is oriented to questions of toxicity in livestock; for convenience the toxic plants from these families are grouped according to the chemical classification of their toxic constituents, rather than botanically. The groups of toxins to be considered are shown in table 3.1. Although the toxicity of these plants to livestock will be empha-

Table 3.1 Toxin Classes of Toxic Genera of the Solanaceae and Liliaceae

Toxin Class	Genera	Family
Nonalkaloids	*Cestrum*	Solanaceae
	Solanum	Solanaceae
Tropane alkaloids	*Atropa*	Solanaceae
	Datura	Solanaceae
	Hyoscyamus	Solanaceae
Pyridine/piperidine alkaloids	*Nicotiana*	Solanaceae
Steroidal alkaloids	*Veratrum*	Liliaceae
	Schoenocaulon	Liliaceae
	Zygadenus	Liliaceae
	Lycopersicon	Solanaceae
	Solanum	Solanaceae

sized, the demonstrated teratogenic effects of some species on laboratory animals indicates their potentially severe hazards to humans.

Supported in part by an interagency agreement between the U.S. Department of Agriculture, Agricultural Research Service, and the National Institutes of Health, National Institute of General Medical Sciences, as funded by the latter.

Economic Aspects of Poisonous Plants in Animal Agriculture

The 1.2 billion acres of native forest and rangelands represent about 63 percent of the total land area of the 48 contiguous United States. About two-thirds of the forest and rangelands are grazed annually by livestock. Excluding timber and mineral values, the primary production value of those forest and rangelands is for livestock grazing (U.S. Dept. of Agriculture, 1974).

The production value of 8 of the principal cash crops of the 48 contiguous states and the acres harvested (U.S. Dept. of Agriculture 1974, 1976) in 1970 is shown in table 3.2. Crops are listed in ascending order of

Table 3.2 Production Values of Various Agricultural Commodities in the United States (1970)

Cash Crops	Acres Harvested (Millions)	Production Value (Billions of Dollars)
Peanuts	1.5	0.38
Sugar beets	1.4	0.39
Potatoes	1.4	0.72
Tobacco	0.9	1.39
Wheat	43.0	1.80
Forage from native grazing lands[a]	835.0	2.55[b]
Soybeans	42.0	3.20
Corn	57.0	5.50

[a]All forest and range grazing lands in the United States (public and private).
[b]Calculated from value of 213 million animal unit months $\simeq$ yearly forage requirements for 17 million cows $\simeq$ 17 × \$300/2 = \$2.55 billion/year. Taken from USDA agricultural statistics.

production value in billions of dollars. The harvest from grazing lands had a very high production value. The range had a capacity of 213 million animal unit months. The grazing land harvest, therefore, had a gross value of \$2.55 billion in 1970 (derived from a harvest value of \$300 for each of about 17 million cattle grazing for 24 months). The net return would have been much higher had animals not grazed poisonous plants. It is impossible to determine accurately the dollars lost from this grazing because no reporting system exists. But from our experience and the experience of those who preceded us in poisonous plant studies in the U.S. Department of Agriculture, we believe that about 5 percent of all grazing livestock are seriously affected by poisonous plants each year. The nature of the losses from such encounters is variable. The effects can be either direct, causing death, debilitation, reduced weight gain, or management problems in the grazing animal, or indirect, causing abortions or congenital deformities in the unborn offspring of that animal. Of the 5 percent of grazing animals that are affected one way or another by poisonous plants, about two-fifths of these (or 2% of all grazing livestock) either die, are debilitated, abort, or have congenitally deformed offspring. That represents an economic loss of about $51 million annually.

Animals ingest poisonous plants for various reasons. They like some of them and can often be found voraciously grazing species of *Delphinium*, *Lupinus*, *Veratrum*, *Halogeton*, and *Astragalus*, to name but a few. Some they evidently find distasteful, but they graze them when other forage is in short supply; these include *Tetradymia*, *Datura*, and *Helenium* species. Often their choices are limited when they find themselves in areas where poisonous plants predominate. Another factor causing problems is when livestock are supplementally fed by ranchers with crop residues that are poisonous. Among the many plants causing livestock problems are members of the Solanaceae and Liliaceae.

Nonalkaloidal Toxins

Several toxic plants of the Solanaceae and Liliaceae contain nonalkaloidal toxins. Three solanaceous plants that are included in this category are *Cestrum parqui*, *Cestrum diurnum*, and *Solanum malacoxylon*.

Cestrum parqui grows wild in coastal plains areas from Florida to Texas. Although not reported to be a serious problem in the United States, it does cause animal poisoning in Australia. The plant has become naturalized in that country and is now a common weed in southeastern Queensland and northeastern New South Wales (Everist 1974). Cardiac glycosides have been isolated from the plant (Canham and Warren 1950) and could well account for the pathologic signs in poisoned animals. These signs include hemorrhages of the heart, lung, and gastrointestinal tract. We have found similar severe hemorrhages in animals experimentally poisoned with certain related cardiac glycosides (Benson et al. 1977) from milkweed plants.

Another member of the *Cestrum* genus, *C. diurnum*, which grows in the southeastern United States, has been implicated as a plant causing calcinosis of the type to be described later under *Solanum malacoxylon*. The condition has been reported in cattle and horses in the Miami area (Wasserman 1975).

The plant *Solanum malacoxylon* produces perhaps the most interesting and certainly the most studied and documented toxicity among the nonalkaloidal toxins of the Solanaceae and Liliaceae (Wasserman 1975; Wasserman et al. 1976; Peterlik et al. 1976; Haussler et al. 1976; Basudde and Humphreys 1976; Proscal et al. 1976; Puche et al. 1976). The plant is native to certain South American countries, where it is grazed by livestock. Animals grazing this plant develop symptoms of calcinosis and pathologic signs similar to hypervitaminosis D (Wasserman 1975). They have excessive circulating calcium and phosphorus and extensive soft tissue calcification. They have stiff limbs and become emaciated, and some animals die (Wasserman 1975; Wasserman et al. 1976). Wasserman and associates (1976) have shown that the toxicity of the plant is due to the glycoside of 1,25-dihydroxyvitamin D_3 (see figure 3.1), which is the biologically active form of vitamin D and is biosynthesized in the kidney. This form of the vitamin me-

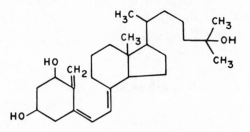

1,25-DIHYDROXYVITAMIN D₃

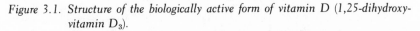

Figure 3.1. Structure of the biologically active form of vitamin D (1,25-dihydroxy-vitamin D_3).

diates calcium and phosphorus absorption from the intestine. Thus, plant ingestion remarkably increases the level of 1,25-dihydroxyvitamin D_3 in the animal and gives rise to excessive absorption of calcium from the intestine and synthesis of the calcium-binding protein.

Wasserman (1975) speculates on the physiologic basis for the effect as follows:

> It is now established that cholecalciferol (vitamin D_3), once acquired by the body, is converted into more polar metabolites that are more biologically active

than vitamin D *per se*. The first important reaction occurs in the liver, in which the vitamin is converted to the 25-hydroxylated derivative [25-(OH)D₃] and subsequently hydroxylated in the 1α-position in the kidney to produce 1α,25-(OH)₂D₃. The 1α,25-(OH)₂D₃ metabolite is the most active of the cholecalciferol derivatives yet identified and, importantly, its rate of formation appears to be directly related to the calcium needs of the animal. As an example, when a calcium-deficient diet is fed, the recipient animal in some fashion can detect this circumstance and increase the amount of 1α,25-(OH)₂D₃ produced by the kidney enzymes. As a consequence, the efficiency of calcium absorption is increased. On the other hand, an intake of a diet containing adequate levels of calcium and phosphorus reduces the amount of 1α,25-(OH)₂D₃ formed and this is followed by a decrease in the intestinal transport mechanism. Thus, there is a feedback regulation of the activity of the 1α-hydroxylase system of the kidney, which, through the controlled production of 1α,25-(OH)₂D₃, attempts to assure either that a sufficient amount of calcium is absorbed to meet physiological needs or that there is not an overabundant absorption of the element. If 1α,25-(OH)₂D₃ (or a compound that acts like this active metabolite) is administered, the control mechanism is circumvented and calcium absorption would proceed at a rate related to the amount of administered 1α,25-(OH)₂D₃ rather than to the calcium needs of the animal. (p. 2)

Tropane Alkaloid Toxins

Among the plants containing tropane alkaloids, *Atropa belladonna* (deadly nightshade), *Datura stramonium* (jimsonweed), and *Hyoscyamus niger* (henbane) are perhaps the most common. *D. stramonium* causes frequent and serious toxicosis in domestic livestock. It is a large, coarse annual with tubular flowers. The plant is distributed throughout the United States, especially in the Southwest (Kingsbury 1964). All *Datura stramonium* plant parts are highly toxic because of high alkaloid concentration. In man the signs of poisoning by the plant vary somewhat with the relative concentration of the various alkaloids and are characteristic of the individual alkaloids (Kingsbury 1964). Thirst, vision disturbance, flushed skin, hyperirritability of the central nervous system, delirium, rapid and weak heartbeat, and convulsions are among the symptoms of intoxication.

Hyoscyamine is the most common alkaloid of the tropane derivatives found in these plants. It is an ester readily hydrolyzed to tropine and tropic acid, two products devoid of the physiologic properties of the parent compound (Swan 1967). The racemic form of hyoscyamine, atropine, is commonly used as a drug in ophthalmology for pupillary dilation. Isolated atropine is probably derived from the ready racemization of hyoscyamine during isolation procedures.

The alkaloids are hallucinogenic and were used by sorcerers in the Middle Ages. The Babylonians used *Hyoscyamus niger* seeds for relief of toothache, and the Nubians smoked *Datura stramonium* leaves for relief of asthma (Swan 1967).

Kingsbury (1964) has reviewed some of the literature on poisonings in livestock. He states that deaths have been reported in all classes of livestock, including horses, cattle, sheep, hogs, mules, and chickens—in the latter case from ingestion of seeds. Animals generally avoid jimsonweed unless food is scarce. Pammell (1911) summarized a number of accidental poisonings in humans that came to his attention in 1897 as follows:

> Cases of poisoning arise in adults from excessive use of a stimulant or a medicine. Children are sometimes tempted to eat the fruit, if they are permitted to play where the weed is to be found. Several cases of this kind were reported to the Department during the fall of 1897. At Alpena, Michigan, five children were badly poisoned in August by eating the seeds of the purple-flowered species, which was cultivated in a garden as a curiosity under the fanciful trade name of "Night-blooming Cactus." In September a boy was killed in New York by eating the seeds of jimsonweed, which was permitted to grow in a vacant lot; his brother, poisoned at the same time, was saved only with difficulty. In October two other cases occurred in New York. Four children were playing in one of the public parks of the city where jimsonweeds were growing luxuriantly. The boys imagined themselves Indians and roamed about and ate parts of various plants. Three of them ate the seeds of the jimsonweed. One died in a state of wild delirium; another was saved after heroic treatment with chloral hydrate and morphine; the third, who ate but few of the seeds, was but little affected. Children are also poisoned by sucking the flower, or playing with it in the mouth. The fresh green leaves and also the root have occasionally been cooked by mistake for other wild edible plants. One or two instances are recorded in which cattle have been poisoned by eating the leaves of young plants which were present in grass hay, but these animals generally either avoid the plants or are very resistant to its poison. (pp. 730–31)

According to the summary by Watt and Breyer-Brandwijk (1962), poisoning of humans by *Datura* spp. in southern Africa is quite common. They report that, in 1943 alone, 1,524 African soldiers were poisoned in a few separate incidents when leaves had been gathered for greens. A similar instance in Jamestown, Virginia, early in U.S. history was the event from which the common name jimsonweed (Jamestown weed) was derived.

Although the toxicity of plants containing tropane alkaloids has been extensively documented because of widespread toxicosis, few reports of any potential teratogenic hazard exists for those plants or alkaloids. Beuker and Platner (1956) injected atropine into fertile chicken eggs during the fourth to twelfth days of incubation, but no defects occurred. However, Leipold, Oehme, and Cook (1973) reported an outbreak of arthrogryposis in 25 offspring from 8 litters of newborn pigs that they speculated was due to maternal ingestion of *D. stramonium*. The pregnant sows were farrowed in a pen surrounded by a dense stand of *D. stramonium* and developed typical signs of *D. stramonium* intoxication during the second and third months of pregnancy; the plant had obviously been foraged. The authors did not believe that genetic inheritance was involved because

cases did not follow an hereditary pattern inasmuch as various age groups of sows were involved, crossbreeding was practiced, and seasonal influence related to growth of jimsonweed was observed. Further, the observed proportion of cases (25 of 84 newborn pigs) was too high to fit a simple recessive gene pattern. The strongest argument against an hereditary pattern is failure to repeat observation of congenitally defective pigs in the next breeding season, using the same breeding groups. (p. 1060)

But unequivocal evidence that *D. stramonium* is responsible must still be obtained by feeding trials. If such evidence is forthcoming, then it will be most interesting to know whether one (or more) of the contained toxic tropane alkaloids is also a teratogen.

Pyridine/Piperidine Alkaloidal Toxins

Although the tropane alkaloids are of rather widespread distribution among solanaceous species, the pyridine/piperidine alkaloids are found almost exclusively in the *Nicotiana* genus, with a few in *Duboisia* and *Withania* (Bóttomely and White 1951; Majumdar 1952; Cosson, Vaillant, and Dequeant 1976). Essentially absent among members of the Liliaceae, pyridine/piperidine alkaloids are widespread among families other than the Solanaceae.

Among the most commercially important plants in the world, *Nicotiana tabacum* (tobacco) is also among the most hazardous. Its toxicity is advertised widely. Every pack of cigarettes in the United States carries the inscription "Warning: The Surgeon General has determined that cigarette smoking is dangerous to your health.". However, here we consider the toxicity of the constituent alkaloids, excluding pyrolysis products.

A variety of alkaloids have been reported in *Nicotiana tabacum*. Figure 3.2 shows the structures of some of the more common alkaloids. Nicotine, the most abundant alkaloidal constituent, has an N-methylpyrrolidine ring linked by its 2 position to the 3 position of a pyridine ring. Other alkaloids present are derived largely from a pyridine ring to which a functional group, usually a 5- or 6-membered nitrogen-containing ring of varying saturation, is attached.

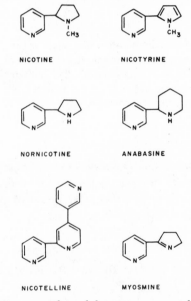

Figure 3.2. *Structure of six of the most common tobacco alkaloids.*

The toxicologic effect of nicotine has been extensively studied and reviewed (Watt and Breyer-Brandwijk 1962; Gorrod and Jenner 1975). The compound is rapidly absorbed in humans through the lungs, mouth, stomach, and skin but is also rapidly excreted in the urine to virtual completeness within about half a day. It is metabolized through a complex series of reactions upon passage through the liver of mammals, and excreted in part as metabolites. The fatal dose in a 50 kg human is near 40 mg orally. At lower doses humans experience nausea and a brief stimulation followed by depression, vomiting, tremors, and convulsions. We find that 50 kg sheep can be dosed orally up to 2 g, and cows can be dosed orally up to 3 g of free base nicotine before showing symptoms of toxicity. The degradation capabilities of sheep and cow rumen or liver, and their excretion abilities, are remarkably more efficient than those of man. Watt and Breyer-Brandwijk (1962) have reviewed many interesting cases of human poisonings from nicotine or *Nicotiana tabacum*. They include the death of a child from the blowing of soap bubbles with a foul tobacco pipe; a variety of deaths from *N. tabacum* decoctions used as enemas; the accidental swallowing of 0.8 g snuff; and the case of a smuggler who, wrapping leaves of the plant about his body to evade customs duty on the product, perspired enough to moisten the leaves and thereby absorb through the skin a lethal dose of nicotine.

The possible teratogenicity of *Nicotiana tabacum* and certain contained alkaloids has been investigated by many workers. Vara and Kinnunen (1951) showed that nicotine causes fetal resorptions. According to Nishimura and Nakai (1958), nicotine when injected (s.c. or i.p.) into pregnant mice caused resorptions, fetal deaths, decreased litter size, and congenital deformities of the skeletal system. The deformities included defects of the elbow and wrist joints, digits, palate, and spine. According to Landauer (1960), injection of nicotine sulfate into fertilized chicken eggs at various stages of incubation and at various dosages resulted in teratogenic effects in hatched chicks. Effects included a shortening and twisting of the neck due to incomplete and irregular formation and fusion of cervical vertebrae, dwarfing, shortening of the upper beak, and muscular hypoplasia.

Crowe (1969) described observations made during 1967–1969 of more than 300 skeletally deformed pigs from 64 litters on 5 Kentucky farms. During gestation all the sows had access to *Nicotiana tabacum* stalks. He speculated that the stalks were responsible for the deformities. Following that report Menges and his co-workers (1969) described a similar *N. tabacum*-related epidemic of congenital deformities in swine. Twisting of the fore or hind limbs and dorsal flexure of hind limb digits were the principal manifestations. Fourteen out of 79 sows and gilts with access to *N. tabacum* stalks during essentially the first 40 days of gestation delivered 149 offspring; 40 percent of these were malformed. Dams not exposed to the stalks as a food source during gestation had no offspring with birth defects. The 14 dams that gave birth to malformed offspring during the epidemic had no malformed offspring in prior or later litters, despite breeding by the same boars. The investigators speculated that nicotine from the stalks was the most likely cause of the malformations.

Crowe and Pike (1973) reported further observations on this problem from examination of over 1,000 malformed newborn pigs born during a 5-year period. The susceptible period of gestation, they concluded, was between day 10 and day 30. The principal manifestation of the condition was arthrogryposis, with occasional involvement of the spinal column or mandible.

In 1974 Crowe and Swerczek (1974) produced congenital arthrogryposis by feeding aqueous *Nicotiana tabacum* leaf filtrates to pregnant sows. They did not determine whether the alkaloids were responsible, but reported that negative results suggested that nicotine was probably not the active principle.

Feeding trials in sheep and cows with free base nicotine at our laboratory, in cooperation with Dr. Crowe, have not produced unequivocal congenital deformities. We have conducted extensive feeding trials in cows on a number of pyridine/piperidine alkaloids; the results of these trials suggest a possible identity of the *Nicotiana tabacum* teratogen. These trials were conducted because we found that the plant *Conium maculatum* (poison hemlock) and two of its contained piperidine alkaloids were responsible for certain congenital defects that occur in cattle under natural conditions. When pregnant cows ingest the plant during the 50–75 days of gestation, they deliver calves with arthrogryposis, scoliosis, torticollis, cleft palate, or a combination of these. The alkaloids coniine and probably γ-coniceine (see figure 3.3) from the plant fed during that period produce the same effects as the plant (Keeler 1974).

CONIINE   γ-CONICEINE

Figure 3.3. Structure of the piperidine alkaloids coniine and γ-coniceine.

Because of the widespread presence of pyridines and piperidines in plants ingested by animals, we investigated the chemical structural requirements for teratogenicity among these compounds (Keeler and Balls, 1978). To do so, we administered a number of coniine analogs to pregnant cows. Figure 3.4 shows coniine and the analogs tested. None of the seven analogs produced terata when gavaged at three to seven times as high a dose as coniine. The results suggest that both alkyl chain length and degree of unsaturation in the ring are critical. Neither 2-ethylpiperidine nor 2-methylpiperidine

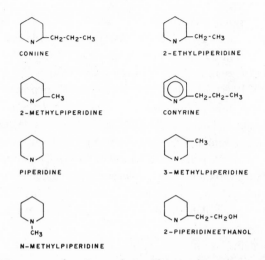

Figure 3.4. Structures of coniine and piperidine analogs administered to cows.

was teratogenic, but coniine itself (2-propylpiperidine) was. Furthermore, although a fully saturated ring (coniine) or apparently one with a single double bond (γ-coniceine) was active, the fully unsaturated ring compound conyrine was not active. Not surprisingly, therefore, piperidine, 3-methylpiperidine, N-methylpiperidine, and 2-piperidineethanol were inactive (Keeler and Balls, 1978). Thus, among known *Nicotiana tabacum* alkaloids, the compound we most strongly expect to be a teratogen would be anabasine because it does have an α-substituted piperidine ring and the α group (pyridine) is certainly larger than propyl, although much more reactive. We are conducting feeding trials to test the hypothesis that anabasine is a teratogen.

Steroidal Alkaloid Toxins

The toxic steroidal alkaloids from members of the Solanaceae and Liliaceae are scientifically and economically important because some have been used in medicinal preparations and as insecticides. For example, preparations of the ester alkaloids of *Veratrum* have been used in the treatment of certain types of hypertension (Krayer 1958), and mixed alkaloidal preparations of *Veratrum* and *Schoenocaulon* were used as insecticides (Kingsbury 1964; Swan 1967). *Zygadenus* and *Veratrum*, both common range genera, are frequently grazed by livestock and are considered poisonous plants. Their toxicity is due to steroidal alkaloids (Kingsbury 1964). Likewise, crop residues from *Solanum tuberosum* (potato) and *Lycopersicon esculentum* (tomato) cause frequent toxicoses in livestock when they are fed these materials (Kingsbury 1964). Finally, *S. tuberosum* and *L. esculentum* and certain other members of the Solanaceae, e.g., *Solanum melongena* (eggplant), as part of the human diet occasionally give rise to toxicoses owing to their content of steroidal alkaloids (Hansen 1925; Nishie, Gumbmann, and Keyl 1971, Patil et al. 1972; Jadhav and Salunkhe 1975). There is a considerable range of structure variation in ring systems of steroidal alkaloids involved in the genera just mentioned (Kupchan and By 1968; Schreiber 1968). Naturally occurring representatives often include glycosides, esters, and the free parent alkamines.

The ring systems (see figure 3.5) in most of the steroidal alkaloids in the *Veratrum*, *Zygadenus*, and *Schoenocaulon* genera, commonly called alkaloids of the veratrum group, are C-nor-D-homo steroids, with an additional 1 or 2 rings. The terminal ring is invariably a piperidine one methylene removed from the steroid. The methylene is attached at C_{17} of the steroid and α to the nitrogen of the piperidine. Many compounds have been isolated in which a methylene bridge is also attached between the nitrogen of the piperidine and carbon 13 of the steroid, giving rise to six-ring alkaloids with

RING SYSTEMS OF VERATRUM ALKALOIDS

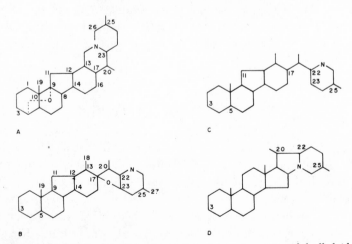

Figure 3.5. Examples of the ring systems in veratrum steroidal alkaloids.

tertiary nitrogen atoms (ring system A). Examples include protoverine, germine, veracevine, zygadenine, and a variety of esters. Four compounds have been reported with an ether bridge between the steroid and the piperidine attached at C_{17} of the steroid and α to the nitrogen of the piperidine, giving rise to a furanopiperidine. The furanopiperidine is thus attached spiro at C_{17} so that it is at right angles to the plane of the steroid (ring system B). The compounds are jervine and its glycoside pseudojervine and cyclopamine (11-deoxojervine) (Keeler 1969a) and its glycoside cycloposine (Keeler 1969b). Compounds with only a methylene bridge (C_{17}–C_{22}) between the pyridine and the steroid (as in ring system C) include veratramine, its glyco-

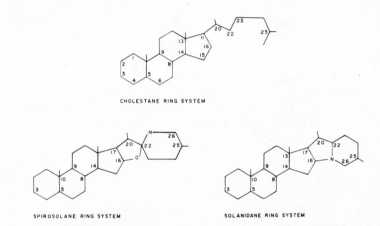

CHOLESTANE RING SYSTEM

SPIROSOLANE RING SYSTEM   SOLANIDANE RING SYSTEM

Figure 3.6. Examples of the ring systems in solanum steroidal alkaloids and their relationship to cholestane.

side veratrosine, and muldamine (Keeler 1971a). Three veratrum alkaloids have the conventional steroid ring system (ring system D) and bear considerable resemblance to the solanidines from *Solanum* species to be described later in this section. Those veratrum alkaloids are rubijervine, isorubijervine, and isorubijervosine, the glycoside of the latter.

Ring systems in alkaloids from members of the *Solanum* and *Lycopersicon* genera (Shreiber 1968; Fieser and Fieser 1959), commonly referred to as the solanum alkaloids (see figure 3.6), invariably have the common C_{27}-carbon steroid skeleton of cholestane. Most, including all to be considered here, have either the spirosolane or the solanidane skeleton. Important examples include solasodine, tomatidine, solanidine, demissidine, and their glycosides.

The pharmacologic and toxicologic effects of steroidal alkaloids vary considerably (Krayer 1958; Nishie, Gumbmann, and Keyl 1971; Patil et al., 1972) as a function of the wide variation in structure of these compounds. The tertiary amine veratrum esters produce a transient decrease in blood pressure, a decrease in arterial resistance, a decrease and irregularity in heart rate, a slowing of the rate of respiration, an emetic action, and certain effects on the neuromuscular system. Salivation and vomiting are the effects we see most frequently from overdose of *Veratrum* in sheep. The secondary amines are much less active on a weight basis, and their effects are somewhat different. The most characteristic toxic effects we see in sheep from the secondary amines, veratramine and jervine, are convulsions. The solanum glycosides from *S. tuberosum*, such as solanine, give rise principally to gastrointestinal disturbances such as nausea, vomition, diarrhea, and hemolytic and hemorrhagic damage to the gastrointestinal tract.

We established a number of years ago that the liliaceous *Veratrum californicum* caused cyclopian and related cephalic malformations in lambs born to pregnant dams that ingested the plant on the fourteenth day of gestation (Binns et al. 1963, 1965). This teratogenic effect, the etiology of which was determined by investigators in our laboratory demonstrating the role of the plant, occurred in epidemics in areas of Idaho (Binns et al. 1963). Hundreds of lambs were deformed from this cause each year. The extreme anatomical deformities included single or double globe cyclopia (see figure 3.7). The lambs had a shortened upper jaw and a protruding lower jaw, sometimes with a peculiar skin-covered proboscis above the single eye. Double globe cyclopia was the deformity from which the common name for the condition, monkey face lambs, was derived. This is the name by which the disease is commonly known among sheep handlers. In examples of mild severity of the disease, animals had normal eyes and only a shortening of the upper jaw or cebocephaly. Our efforts to isolate the active principle indicated that it was likely to be an alkaloid. We fed to pregnant sheep a variety of veratrum alkaloids from the plant to test their teratogenicity. The compounds jervine, cyclopamine, and cycloposine were active and produced deformities typical of natural cases (Keeler and Binns 1968). Other alkaloids were not active.

We established the structural identity of cyclopamine as 11-deoxojervine (Keeler 1969a), a compound nearly identical to jervine (see figure 3.8); the third active compound, cycloposine, was 3-glucosylcylcopamine (Keeler 1969b). Inactive veratramine differs from cyclopamine because it has no oxide bridge forming ring E; it had a hydroxyl function on ring F and a

Figure 3.7. *Double globe congenital cyclopia from maternal ingestion of the plant* Veratrum californicum *on the fourteenth day of gestation.*

greater degree of unsaturation in ring D. Another inactive compound, muldamine (Keeler 1971a), differs from cyclopamine because it has no ether
bridge, no unsaturation in ring D, and an O-acetate function on carbon 11.
We presumed, then, from these differences between active and inactive
compounds that the fused furanopiperidine ring E/F arrangement of cyclo-

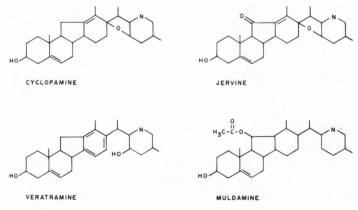

Figure 3.8. *Structural comparison of the teratogenic compounds jervine and cyclopa-*
mine with the nonactive compounds veratramine and muldamine

pamine was essential for the activity (Keeler 1970). In feeding trials in
sheep, only compounds containing that ring arrangement were active.

These teratogens have wide species and anatomical specificity. Fetal insult on day 14 of gestation in sheep produced cyclopia and related cephalic
deformities. Insult on other gestation days produced anatomical deviations
of other types (Keeler 1973), including limb bone shortening and cleft lip
and palate. Sheep are not the only susceptible ruminant animals. Cattle and
goats were affected (Binns, Keeler, and Balls 1972). Although the clinical
manifestations of the malformations vary, chick embryos (Bryden, Perry,
and Keeler 1973), rats (Keeler 1975), hamsters (Keeler 1975), and rabbits
(Keeler 1971b) have also proved to be susceptible when the teratogens are
administered at about the primitive streak/neural plate stage of embryonic
development. The dosages required for production of deformities in sheep
varied from 1.5 g oral (Keeler and Binns 1968) to about 1 mg by intrauterine
injection (Bryden and Keeler 1973). The most convenient assay, and the
one we are currently using in studies on the teratogenicity of structural
analogs of jervine and cyclopamine, is the hamster assay. A single oral dose
of jervine or cyclopamine of about 20 mg on day 7 of gestation produces
deformities in 50 percent or more of the individuals from each litter. In-
traperitoneal injection of 10–20 mg in 2% methyl cellulose also is a suitable route of administration and has proven useful in competitive inhibition
studies.

Various functional group analogs of jervine and cyclopamine were tested
in the hamster assay, and some altered the teratogenicity of the parent molecule. But among them N-methyljervine was just as active as jervine. Thus,
the nitrogen can be either tertiary or secondary without altering activity
(Keeler, Brown, and Young 1976).

Because of the relationship of the structures of the teratogenic veratrum
steroidal alkaloids to structures of a number of solanum alkaloids, we have
had an interest for many years in the possible teratogenicity of solanum
alkaloids. The report by Renwick (1972) hypothesizing that maternal consumption by humans of blighted potatoes was responsible for congenital
spina bifida and anencephaly has served to stimulate further our interest in
these structural relationships. If there is a teratogen in potatoes, as Renwick
alleged, we wondered whether it could be of the steroidal alkaloid class.

Other scientists have been stimulated to examine the same question.
Swinyard and Chaube (1973) injected pure solanine and extracted potato
alkaloids into pregnant rats and rabbits and produced no neural tube defects.

In a further report Chaube and Swinyard (1976) found no neural tube defects in young from pregnant rats injected i.p. with α-solanine, although fetal deaths resulted. Nishie, Norred, and Swain (1975) reported that α-chaconine, α-solanine, and tomatine failed to produce significant terata in chicks, while Bell and associates (1976) found no terata in offspring from pregnant mice injected i.p. with solanine. In addition, Mun and coworkers (1975) reported that both pure solanine and glycoalkaloids extracted from *Phytophthora*-infected potatoes produced deformities in chick embryos from eggs treated during early development. Jelínek, Kyzlink, and Blattny (1976) verified these observations and in further experimental work reported that

> the embryotoxic factor in ethanol extracts of boiled potatoes infected with *P. infestans* produced the caudal regression syndrome and myeloschisis at somite stages of chicken embryos. After an injection on the third and fourth days a malformation syndrome consisting of cranioschisis, celosoma, and cardiac septal defects was a characteristic consequence. The same syndrome could be induced by injecting an equivalent amount of extract of healthy potatoes and by injecting solanine in amounts corresponding to the solanine concentration in the given extracts. (p. 340)

But our interests centered on alkaloids more closely related to those from *Veratrum*. No known potato alkaloids have a fused furanopiperidine ring E/F arrangement like those from *Veratrum*, but the spirosolane solanum alkaloids have a furan ring E that is fused to the steroidal end of the molecule that in some cases confers a similar configuration upon them. Consequently, we tested the teratogenicity of the configurationally similar solanum alkaloid solasodine and the teratogenic veratrum alkaloids, jervine and cyclopamine, in our hamster assay (Keeler, Young, and Brown 1976). Solasodine has not to our knowledge been reported to be present in potatoes, although its glycosides are common in eggplant, but solasodine is configurationally closer than some potato constituents. Also examined were tomatidine (configurationally and otherwise nearly identical to Δ^5-tomatidine-3β-ol, the aglycone of certain minor glycosides found in some potatoes, but with a piperidine ring configuration opposite that of solasodine), jervine, and cyclopamine. Solasodine, jervine, and cyclopamine were compared with diosgenin, the non-nitrogen-containing analog of solasodine.

The results (Keeler, Young, and Brown 1976) established that solasodine was teratogenic (see figure 3.9). Deformities in hamster fetuses resulted when dams were gavaged 1200 to 1600 mg/kg (doses nearly one order of

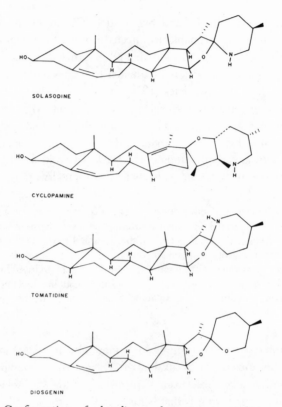

Figure 3.9. Configurations of solasodine, cyclopamine, tomatidine, and diosgenin.

magnitude higher than the hamster teratogenic doses of cyclopamine). The solasodine analogs tomatidine and diosgenin were not teratogenic at doses double that of solasodine. The essentiality of the nitrogen or piperidine ring is suggested by the lack of teratogenicity of diosgenin, which is nitrogen free but otherwise identical to solasodine. The importance of the configuration of the nitrogen-containing ring is suggested by lack of activity of tomatidine, a close analog of solasodine with opposite or β-piperidine ring configuration. Results from preliminary experiments (Brown and Keeler 1977) suggested that saturation at the 5–6 bond, as in tomatidine (the only difference aside from the piperidine ring difference between tomatidine and solasodine), was not necessary for teratogenicity. Although the C-nor-D-homo ring system is not essential for teratogenicity, nitrogen configuration is important. While the nitrogen in solasodine, like cyclopamine, is on the α side of the molecule, in cyclopamine it extends much more markedly in the α direction, suggesting that the extent of projection in this α direction is important.

Consequently, among solanum alkaloids the spirosolanes based on tomatidine or Δ^5-tomatidine-3β-ol common to potatoes and other food sources with nitrogen on the β side of the molecule would probably not be active in hamsters (Keeler, Young, and Brown 1976). However, those based on solasodine, none of which apparently is found in potatoes, or any as yet unidentified spirosolanes with α piperidine ring configuration might be active.

The solanum alkaloidal aglycone most common to potatoes is solanidine, a compound considerably different from solasodine. Solanidine (Schreiber

Table 3.3   Susceptibility of Pregnant Hamsters[a] to Potato Sprouts and Steroidal Alkaloids

	Total no. of Pregnant Litters[b]	% Deformed Litters	Statistical Significance[c]	% Totally Resorbed Litters	% Dams That Died from Overdose
H$_2$O Controls	522	1.34		1.34	0
Kennebec sprouts	181	25.40	P<.0005	9.40	29.3
Russet sprouts	70	25.70	P<.0005	17.00	39.7
Pioneer sprouts	60	8.30	P<.0005	0	16.7
Targhee sprouts	58	10.40	P<.0005	0	22.7
Sebago sprouts	52	13.50	P<.0005	9.60	48.5
Nampa sprouts	55	18.20	P<.0005	9.10	17.9
Norchip sprouts	46	8.70	P≈.0005	0	11.5
Kennebec peel (sprouted)	96	3.13	P>0.2	3.13	20.7
Kennebec peel (unsprouted)	64	4.69	P>0.3	3.13	30.4
Kennebec tuber (sprouted)	124	1.61	P>0.8	0.81	4.6
Kennebec tuber (unsprouted)	72	1.38	P>0.9	0	7.7
Jervine	26	73.10	P<.0005	3.80	7.1
Cyclopamine	17	47.10	P<.0005	5.90	51.4
Solasodine	105	26.70	P<.0005	20.00	8.7

[a] Simonsen Laboratories—pre-1976 strain.
[b] Carried to term or resorbed.
[c] Statistical significance of the difference between the incidence of malformations in control and treated groups.

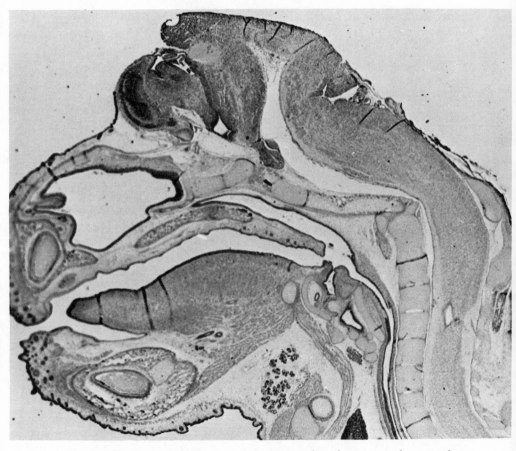

*Figure 3.10. Example of an exencephalic hamster fetus from maternal gavage of po-
tato sprouts.*

1968) is devoid of a furan ring E, with no spiro connection between rings D
and E or between rings E and F; therefore, the tertiary nitrogen of solanidine is not α to the plane of the steroid as a consequence of a piperidine ring
at right angles to the steroid system.

A colleague, Dr. Dennis Brown, believed that steroidal alkaloids with a
tertiary nitrogen whose lone electron pair was constrained toward the α side
of the molecule might possibly be active. Consequently, he synthesized
solanidanine isomers at C-22 and C-25. We tested them in the hamster
assay, and, as he predicted, the 22S, 25R solanidanines, whose lone ni-

trogen electron pair is constrained α to the molecule, were active, while the 22R, 25S solanidanines, with a β-constrained electron pair, were not (Brown and Keeler 1978). Solanidanines with a β-constrained electron pair on the nitrogen are the naturally occurring isomers.

We have found (Keeler et al. 1976; Keeler et al. 1978) high-alkaloid potato sprouts, but not tubers or peel from tubers, from a number of potato cultivars to be teratogenic in hamster assays in a particularly susceptible hamster strain (see table 3.3). The terata included spina bifida and exencephaly (see figure 3.10). We have not yet identified the teratogen. But on the basis of the foregoing information either a spirosolane or a solanidane with a rigidly positioned nitrogen whose lone electron pair on the nitrogen projected in the α direction seem to be good possibilities.

REFERENCES

Basudde, C. and D. Humphreys. 1976. "The Vitamin D_3 Metabolite-type Activity of *Solanum malacoxylon.*" *Clin. Endocrinol. (Oxford) Suppl.* 5:109–19s.

Bell, D., J. Gibson, A. McCarroll, G. McClean, and A. Geraldine. 1976. "Embryotoxicity of Solanine and Aspirin in Mice." *J. Reprod. Fertil.* 46:251–59.

Benson, J., J. Seiber, R. Keeler, and A. Johnson. 1977. "Comparative Toxicology of Cardiac Glycosides from the Milkweed *Asclepias eriocarpa.*" Society of Toxicology Abstracts, pp. 1–2.

Beuker, E. and W. Platner. 1956. "Effect of Cholinergic Drugs on Development of Chick Embryo." *Proc. Soc. Expt. Biol. Med.* 91:539–43.

Binns, W., L. James, J. Shupe, and G. Everett. 1963. "A Congenital Cyclopiantype Malformation in Lambs Induced by Maternal Ingestion of a Range Plant, *Veratrum californicum.*" *Am. J. Vet. Res.* 24:1164–74.

Binns, W., J. Shupe, R. Keeler, and L. James. 1965. "Chronologic Evaluation of Teratogenicity in Sheep Fed *Veratrum californicum.*" *J. Am. Vet. Med. Assoc.* 147:839–42.

Binns, W., R. Keeler, and L. Balls. 1972. "Congenital Deformities in Lambs, Calves, and Goats Resulting from Maternal Ingestion of *Veratrum californicum*: Hare Lip, Cleft Palate, Ataxia, and Hypoplasia of Metacarpal and Metatarsal Bones." *Clin. Toxicol.* 5:245–61.

Bottomely, W. and D. White. 1951. "The Chemistry of Western Australia Plants. 4. *Duboisia hopwoodii.*" *Aust. J. Sci. Res.* 4:107–11.

Brown, D. and R. Keeler. 1977. "Structure–Activity Relation of Steroidal Amines– Teratogenic Solanidine Derivatives." *Am. Chem. Soc. Abstracts MEDI* 86.

—— 1978. "Structure-Activity Relation of Steroid Teratogens. 3. Solanidan epimers." *J. Agric. Food Chem.* 26:566–69.

Bryden, M. and R. Keeler. 1973. "Effects of Alkaloids of Veratrum californicum on Developing Embryos." J. Anat. 116:464.

Bryden, M., C. Perry, and R. Keeler. 1973. "Effects of Alkaloids of Veratrum californicum on the Developing Chick Embryo." Teratology 8:19–25.

Canham, P. and F. Warren. 1950. "Saponins 2. Isolation of Gitogenin and Digitogenin from Cestrum parqui." J. S. Afr. Chem. Inst. 3:63–65.

Chaube, S. and C. Swinyard. 1976. "Teratological and Toxicological Studies of Alkaloidal and Phenolic Compounds from Solanum tuberosum." Toxicol. Appl. Pharmacol. 36:227–37.

Cosson, L., J. Vaillant, and E. Dequeant. 1976. "The Tropane Alkaloids from the Leaves of Duboisia myoporoides from New Caledonia." Phytochemistry 15:818–20.

Crowe, M. 1969. "Skeletal Anomalies in Pigs Associated with Tobacco." Mod. Vet. Pract. (December), 69:54–55.

Crowe, M. and H. Pike. 1973. "Congenital Arthrogryposis Associated with Ingestion of Tobacco Stalks by Pregnant Sows." J. Am. Vet. Med. Assoc. 162:453–55.

Crowe, M. and T. Swerczek. 1974. "Congenital Arthrogryposis in Offspring of Sows Fed Tobacco (Nicotiana tabacum)." Am. J. Vet. Res. 35:1071–73.

Everist, S. 1974. Poisonous Plants of Australia. Cremorne Jct., N.S.W. Australia, Angus and Robertston.

Fieser, L. and M. Fieser. 1959. Steroids, p. 847. New York: Rheinhold.

Gorrod, J. and P. Jenner. 1975. "The Metabolism of Tobacco Alkaloids" In W. Hayes, Jr., ed., Essays in Toxicology, 6, 35–79. New York: Academic Press.

Hansen, A. 1925. "Two Fatal Cases of Potato Poisoning." Science 61:340–41.

Haussler, M., R. Wasserman, T. McCain, M. Peterlik, K. Bursac, and M. Hughes. 1976. "1,25-Dihydroxyvitamin D_3-Glycoside; Identification of a Calcinogenic Principle of Solanum malacoxylon." Life Sci. 18:1049–56.

Jadhav, S. and D. Salunkhe. 1975. "Formation and Control of Chlorophyll and Glycoalkaloids in Tubers of Solanum tuberosum L. and Evaluation of Glycoalkaloid Toxicity." In C. Chichester, ed., Advances in Food Research, 21, 307–54. New York: Academic Press.

Jelínek, R., V. Kyzlink, and C. Blattriy, Jr. 1976. "An Evaluation of the Embryotoxic Effects of Blighted Potatoes on Chicken Embryos." Teratology 14:335–42.

Keeler, R. and W. Binns. 1968. "Teratogenic Compounds of Veratrum californicum (Durand). 5. Comparison of Cyclopian effects of Steroidal Alkaloids from the Plant and Structurally Related Compounds from Other Sources." Teratology 1:5–10.

Keeler, R. 1969a. "Teratogenic Compounds of Veratrum californicum (Durand). 7. The Structure of the Glycosidic Alkaloid Cycloposine." Steroids 13:579–88.

—— 1969b. "Teratogenic Compounds of Veratrum californicum (Durand). 6. The Structure of Cyclopamine." Phytochemistry 8:223–25.

—— 1970. "Teratogenic Compounds in Veratrum californicum (Durand). 9. Structure–Activity Relation." Teratology 3:169–73.

—— 1971a. "Teratogenic Compounds of Veratrum californicum (Durand). 13. Structure of Muldamine." Steroids 18:741–52.

—— 1971b. "Teratogenic Compounds of Veratrum californicum (Durand). 11. Gestational Chronology and Compound Specificity in Rabbits." Proc. Soc. Expt. Biol. Med. 136:1174–89.

—— 1973. "Teratogenic Compounds of Veratrum californicum (Durand). 14. Limb Deformities Produced by Cyclopamine." Proc. Soc. Expt. Biol. Med. 142:1287–89.

—— 1974. "Coniine, a Teratogenic Principle from Conium maculatum Producing Congenital Malformations in Calves." Clin. Toxicol. 7:195–206.

—— 1975. "Teratogenic Effects of Cyclopamine and Jervine in Rats, Mice, and Hamsters." Proc. Soc. Expt. Biol. Med. 149:302–6.

Keeler, R., D. Brown, and S. Young. 1976. "Teratogenicity of Veratrum and Solanum Steroidal Alkaloids and Analogs in Hamsters." Teratology 14:115.

Keeler, R., D. Brown, D. Douglas, G. Stallknecht, and S. Young. 1976. "Teratogenicity of the Solanum Alkaloid Solasodine and of 'Kennebec' Potato Sprouts in Hamsters." Bull. Environ. Contam. Toxicol. 15:522–24.

Keeler, R., S. Young, and D. Brown. 1976. "Spina Bifida, Exencephaly, and Cranial Bleb Produced in Hamsters by the Solanum Alkaloid Solasodine." Res. Commun. Chem. Pathol. Pharmacol. 13:723–30.

Keeler, R. and L. Balls. 1978. "Teratogenic Effects in cattle of Conium maculatum and Conium Alkaloids and Analogs." Clin. Toxicol. 12:49–64.

Keeler, R., S. Young, D. Brown, G. Stallknecht, and D. Douglas. 1978. "Congenital Deformities Produced in Hamsters by Potato Sprouts." Teratology 17:327–34.

Kingsbury, J. 1964. Poisonous Plants of the United States and Canada. Englewood Cliffs, N.J.: Prentice-Hall.

Krayer, O. 1958. "Veratrum Alkaloids." In V. Drill, ed., Pharmacology in Medicine, 2d ed., p. 515. New York: McGraw-Hill.

Kupchan, S. and A. By. 1968. "Steroidal Alkaloids: The Veratrum Group." In R. Manske, ed., The Alkaloids, 10, 193–286. New York and London: Academic Press.

Landauer, W. 1960. "Nicotine-Induced Malformations in Chicken Embryos and Their Bearing on the Phenocopy Problem." J. Exp. Zool. 143:107–22.

Leipold, H., F. Oehme, and J. Cook. 1973. "Congenital Arthrogryposis Associated with Ingestion of Jimsonweed by Pregnant Sows." J. Am. Vet. Med. Assoc. 162:1059–60.

Majumdar, D. 1952. "Alkaloidal Constituents of Withania somnifera." Curr. Sci. 21:46.

Menges, R., L. Selby, C. Marienfeld, W. Aue, and D. Greer. 1970. "A Tobacco Related Epidemic of Congenital Limb Deformities in Swine." Environ. Res. 3:285–302.

Mun, A., E. Braden, J. Wilson, and J. Hogan. 1975. "Teratogenic Effects in Early Chick Embryos of Solanine and Glycoalkaloids from Potatoes Infected with Late-Blight, Phytophthora infestans." Teratology 11:73–76.

Nishie, K., M. Gumbmann, and A. Keyl. 1971. "Pharmacology of Solanine." Toxicol. Appl. Pharmacol. 19:81–92.

Nishie, K., W. Norred, and A. Swain. 1975. "Pharmacology and Toxicology of Chaconine and Tomatine." Res. Commun. Chem. Pathol. Pharmacol. 12:657–68.

82 R. F. Keeler

Nishimura, H. and K. Nakai. 1958. "Developmental Anomalies in Offspring of Pregnant Mice Treated with Nicotine." *Science* 127:877–78.

Pammell, L. 1911. *A Manual of Poisonous Plants*. Cedar Rapids, Iowa: Torch Press.

Patil, B., R. Sharma, D. Salunkhe, and K. Salunkhe. 1972. "Evaluation of Solanine Toxicity." *Food Cosmet. Toxicol.* 10:395–98.

Peterlik, M., K. Bursac, M. Haussler, M. Hughes, and R. Wasserman. 1976. "Further Evidence for the 1,25-Dihydroxyvitamin D-like Activity of *Solanum malacoxylon*." *Biophys. Res. Commun.* 70:797–804.

Proscal, D., H. Henry, T. Hendriakson, and A. Normal. 1976. "1α,25-Dihydroxyvitamin D_3-like Component Present in the Plant *Solanum glaucophyllum*." *Endocrinology* 99:437–44.

Puche, R., M. Locatto, J. Feretti, M. Fernandez, M. Orsatti, and J. Valenti. 1976. "The Effects of Long Term Feeding of *Solanum glaucophyllum* to Growing Rats on Calcium, Magnesium, Phosphorus and Bone Metabolism." *Calcif. Tissue Res.* 20:105–119.

Renwick, J. 1972. "Hypothesis: Anencephaly and Spina Bifida Are Usually Preventable by Avoidance of a Specific but Unidentified Substance Present in Certain Potato Tubers." *Br. J. Prev. Soc. Med.* 26:67–88.

Schreiber, K. 1968. "Steroidal Alkaloids: The *Solanum* group." In R. Manske, ed., *The Alkaloids*, 10, 1–192. New York and London: Academic Press.

Swan, G. 1967. *An Introduction to the Alkaloids*, New York: John Wiley and Sons.

Swinyard, C. and S. Chaube. 1973. "Are Potatoes Teratogenic for Experimental Animals?" *Teratology* 8:349–57.

U.S. Department of Agriculture. 1974. *Opportunities to Increase Red Meat Production from Ranges of the United States*. Washington, D.C.: U.S. Dept. Agr.

—— 1976. *Agricultural Statistics*. Washington, D.C.: U.S. Dept. Agr.

Vara, P. and O. Kinnunen. 1951. "The Effect of Nicotine on the Female Rabbit and Developing Foetus." *Ann. Med. Exp. Biol. Fenn.* 29:202–13.

Wasserman, R. 1975. "Active Vitamin D-like Substances in *Solanum malacoxylon* and Other Calcinogenic Plants." *Nutr. Rev.* 33:1–5.

Wasserman, R., J. Henion, M. Haussler, and T. McCain. 1976. "Calcinogenic Factor in *Solanum malacoxylon*: Evidence That It Is 1,25-Dihydroxyvitamin D_3-Glycoside." *Science* 194:853–55.

Watt, J. and M. Breyer-Brandwijk. 1962. *The Medicinal and Poisonous Plants of Southern and Eastern Africa*. London: E. & S. Livingstone.

4 Pokeweed and Other Lymphocyte Mitogens

Alexander McPherson

Although it has been approximately 50 years since the initial observation that certain plant seed proteins (lectins) have the ability to agglutinate erythrocytes from animals, it was less than 20 years ago that Nowell discovered that among the properties (see table 4.1) of certain of these plant lectins

Table 4.1   Some of the Properties of Various Lectins That Have Been Purified

1. Binding of sugars (D-mannose, D-galactose, N-acetyl-D-glucosamine, N-acetyl-D-galactosamine)
2. Cross-linking and precipitation of polysaccharides and glycoproteins
3. Agglutination of erythrocytes, leucocytes, lymphocytes, tumor cells, microorganisms, and viruses
4. Eucaryotic cell toxins
5. Inhibition of phagocytosis by granulocytes
6. Inhibition of fertilization of ovum by sperm
7. Insulin-like action on fat cells
8. Inhibition of fungal growth
9. Inhibition of fusion by myoblasts (muscle cells)
10. Binding of transition state metal ions
11. Stimulation of interferon production

is the capacity to release the developmental potential of specific resting or dormant cells in the circulatory system of man and transform them to an active growth state (Nowell 1960). This subgroup of the lectins has been termed *mitogens*, and the class of cells affected is primarily that of the immune responsive cells, or lymphocytes. The lymphocytes, which are normally in the quiescent stage (i.e., presumably in a prolonged G_1 or G_0 stage of their cell cycle) are stimulated into an active metabolic state. Once activated, these cells undergo what appears to be a normal sequence of events that includes blastogenesis and culminates in mitosis. The ability to induce a mi-

Table 4.2 A Selection of Mitogens from Various Classes Found to be Useful in Lymphocyte Stimulation

1. Plant Extracts
   Phaseolus vulgaris (phytohemagglutinin)
   Canavalia ensiformis (concanavalin A)
   Phytolacca americana (pokeweed mitogen)
   Lens culinaris (lentil)
   Vicia faba (favin)
   Pisum sativum (pea)
   Wisteria floribunda
   Abrus precatorius
   Hura crepitans

2. Bacterial Products
   Lipopolysaccharide (lipid A)
   Staphylococcal enterotoxin B
   Tuberculin, purified protein derivative (PPD)
   Streptolysin S

3. Antibody Reagents
   Antiimmunoglobulin sera
   Antilymphocyte sera
   Carbohydrate-specific antibodies
   Anti-α_2-macroglobulin
   Anti-α_2-microglobulin

4. Miscellaneous Chemicals
   Sodium metaperiodate
   Phorbol esters (TPA)
   Ca^{++}-ionophore A-23187
   Metal ions (zinc, mercury, nickel)
   Proteolytic enzymes (trypsin, papain)

togenic respònse in lymphocytes has served as a useful model for studying the biochemistry of lymphocyte development and has contributed substantially to our general understanding of the mammalian cell cycle. Furthermore, the similarities in many of the events that follow antigenic and mitogenic stimulation of lymphocytes suggest that the information attained in studies utilizing plant mitogens are germane to our understanding of the events in the immune response. Mitogenic stimulation of lymphocytes may in fact be a prerequisite for obtaining a productive immune response. The use of plant mitogens has thus become a useful tool in the determination of immune responsiveness under clinical conditions and has been employed in the diagnosis and characterization of some pathological states in humans, Hodgkin's disease being one of the first (Hersh and Oppenheim 1965). Although expectations were initially very high for the routine clinical use of plant mitogens, subsequent testing has been somewhat disappointing. In part, this derives from the lack of purification and insufficient characterization of many of the mitogens currently in use with regard to both physicochemical properties and physiological effects.

Plants are not the only source of lectins that act as lymphocyte mitogens; others are shown in table 4.2. Lectins have the advantage that they are relatively easy to obtain in quantities that permit extensive biological and chemical characterization (Sharon and Lis 1972) and, in addition, show a wide range of specificities. That is, they select only certain classes of carbohydrate receptors on lymphocytic cell surfaces, which then mediate the mitogenic responses. It should be emphasized, however, that of the plant lectins only relatively few are potent lymphocyte mitogens.

The Action of Plant Lectins

The mitogenic activity of plant lectins apparently resides in their ability to bind to and interact with carbohydrate receptors in the cell plasma membrane that are presumably polysaccharide or glycoprotein in nature (Andersson et al. 1972; Greaves and Bauminger 1972). There are several lines of evidence that indicate that the mitogenic stimulus is dependent on the cross-linking of these membrane receptors, an effect that derives from the multivalent character of the oligomeric protein lectins. The earliest events in mitogenesis following the binding of the lectins to the cell surface receptors can be monitored by using fluorescent-labeled protein for UV microscopy or ferritin-labeled lectin for electron microscopy. In either case, time course studies show that frequently the lectin receptor complexes congregate in small areas or patches on the membrane; these patches then aggregate to form larger domains; and finally the labeled complexes migrate over the surface membrane and accumulate at one pole of the cell. This phenomenon is termed *capping* (Greaves and Janossy 1972; Taylor et al. 1971). The lectin aggregate, or cap, is eventually internalized via pinocytosis, although it seems evident at this point that mitogenic stimulation is elicited by interaction with the cell surface and is not dependent on the lectin's reaching the interior. In addition to the very prominent capping phenomenon, a number

of other effects have been observed to occur shortly after the binding of mitogen to the cell surface. The lectin from *Canavalia ensiformis*, concanavalin
A, has been found to be capable of activating basophils and mast cells to
release histamine (Siraganian and Siraganian 1974, 1975). An increased
level of GMP is found within a half-hour after mitogen binding to lymphocytes, along with increased uptake of certain metabolites and ions, especially
Ca^{++} (Greaves and Janossy 1972), and an increased metabolism of phospho-
lipids, particularly the turnover of phosphatidylinositol (Fisher and Mueller
1971).

Experiments by Edelman's group with colchicine, an inhibitor of micro-
tubule formation, suggests that mitogenic stimulation may depend on an an-
chorage system consisting of microfilaments and microtubules within the
cell that are joined directly or indirectly to the surface receptors. The microtubules would, therefore, be responsible for the mobility of lymphocyte surface receptors and the modulation of receptor movement. Edelman further
suggests that such a system may also serve the function of regulating and
transducing a variety of external signals to the appropriate intracellular mes-
sengers (Edelman, Yahara, and Wang 1973).

The study of mitogenic stimulation and the immune response has been
both complicated and enriched by the finding that the general lymphocyte
population is itself composed of a number of subgroups, each of which can
be described in terms of specific cellular responses, physiological responsi-
bilities, or cell structure. The most prominent of these are the thymus-
dependent small lymphocytes, or T cells, and the bone marrow-dependent
B cells. The former are primarily responsible for the cell-mediated immune
response, while the latter are chiefly involved in the elaboration of specific
immunoglobulins (Claman 1970). Although these two types of lymphocytes
can be further divided on the basis of specific structural or functional properties, it is meaningful to consider them at this level as essentially homoge-
neous classes. These two lines of cells differ considerably in their mitogenic
response to the various agents seen in table 4.2. While nearly all of the
known plant mitogens stimulate T cells to enter blastogenesis and division,
very few cause this response in B cells. Thus, for example, when spleen
lymphocytes from athymic nu/nu mice, which presumably possess only B
lymphocytes, are tested for incorporation of tritiated thymidine into DNA
following treatment with plant mitogens such as phytohemagglutinin, PHA,
little or no uptake is observed. The heterozygous litter mates, which have a

normal thymus and both lymphocyte populations present, show normal stimulation.

Pokeweed Lectins

Among the very few plant lectins that have the capability of activating both T and B lymphocytes is that known as pokeweed mitogen, a crude extract of the roots of *Phytolacca americana*. Attempts have been made to

Table 4.3 Molecular Weight in Daltons and Concentration of Pokeweed Mitogens for Optimal Stimulation of Lymphocytes

Mitogen	Molecular Weight	Concentration for Optimal Stimulation (μg/ml)
PA-1	22,000	10–100
PA-2	31,000	1–100
PA-3	25,000	10–100
PA-4	21,000	50–1000
PA-5	19,000	50–500

Source: From Waxdal, 1974.

isolate the active principle from these crude extracts using both classic purification techniques and, more recently, affinity chromatography. Reisfield and associates (1967) reported the isolation of a protein having a molecular weight of 31,000 daltons that possessed potent mitogenic capabilities but did not show stimulation of B cells when tested on athymic mouse spleen lymphocytes. Subsequent to this work Waxdal (1974) was able to purify a set of five glycoproteins, termed Pa^{-1} through Pa^{-5}, from crude pokeweed extract; all of them exhibited some form of mitogenic capacity. The molecular weights of these proteins are shown in table 4.3. Of the five proteins, Pa^{-1} was the most potent mitogen for mixed spleen lymphocyte cultures from mice and was also found to be responsible for all of the specific stimulation of B lymphocytes.

The lectin Pa^{-1} was different from the other mitogens in that it existed as a polymer of 22,000-dalton protomers while the others were observed only in monomeric form. Pa^{-2} through Pa^{-5} have rather unique amino acid compositions, of which asparagine, glutamine, and glycine account for nearly 60 percent. In addition, they contain an unusually large number of disulphide bonds, 20 to 25 in number. The most abundant was Pa^{-2}, which is strictly T cell specific, as are Pa^{-3} through Pa^{-5}. Pa^{-2} has a molecular weight of 31,000 and very likely corresponds to the mitogen isolated by Reisfield and associates (1967). Pa^{-3} is found in very low amounts and is similar to Pa^{-2}, while Pa^{-4} and Pa^{-5} resemble each other in their properties (Waxdal and Basham 1974; Basham and Waxdal 1975; Waxdal, Nilsson, and Basham 1976). Although Waxdal and colleagues (1976) suggest that the individual proteins are entirely unique and that the possibility of a common precursor is inconsistent with certain combinations in the amino acid compositions, this may not in fact be the case. Because of the considerable similarities among amino acid compositions and the fact that all of the mitogens in pokeweed share common receptor specificities, it in fact seems very possible that all or several of the mitogens are derived from a single, higher-molecular-weight species. Thus, Pa^{-1} through Pa^{-5} may be formed by multiple types of proteolytic cleavages along with different degrees of carbohydrate addition or modification.

It is also interesting that Pa^{-1}, Pa^{-2}, and Pa^{-3} are hemagglutinating while Pa^{-4} and Pa^{-5} are not (Waxdal, Nilsson, and Basham 1976). This indicates that plants that contain no hemagglutinating components may still possess mitogenic proteins whose presence is less obvious than in cases in which mitogenic lectins are found. Also of interest in this regard is the observation that the relative amounts of each of the five mitogenic proteins is not constant but varies considerably, depending on the time of year, growth conditions, and the developmental stage of the plant (Waxdal and Basham, 1974).

Using affinity chromatography techniques based on the coupling of each of the five mitogenic fractions to cyanogen bromide-activated Sepharose, Waxdal was able to isolate and identify the major receptor glycoproteins on the surface of both B and T lymphocytes (Waxdal, Nilsson, and Basham 1976). These receptors, which number about six, include, among others, the glycopeptide from immunoglobulin and several of the cellular antigenic determinants. The surprising feature of these results, it seems, is that the receptors bound by each of the mitogenic fractions from both B and T cell extracts are essentially identical. Hence, it is not possible to ascribe the difference in potency of the various fractions to receptor specificities, although it is still possible to explain potency in terms of the relative affinities. More surprisingly, it seems that there is no major difference between the receptors obtained from the cell surfaces of B and T lymphocytes. Either the activation of the T and B lymphocytes depends on the binding of specific receptors that are in such low concentration that they escape detection by the methods employed, or the specific stimulation arises from the manner in which the mitogens interact with the receptors in the context of their surrounding membrane matrix. Hence, we cannot at this time point to any particular factor or group as essential or responsible for the activation of a specific cell population.

It might be noted in passing that the similarity in the receptors for each of the mitogenic fractions is again consistent with the possiblity that all of the pokeweed mitogens may in fact derive from the same precursor protein by specific proteolytic cleavages or polysaccharide modifications.

It is clear from the investigation of the pokeweed mitogens that if one is attempting to describe the physiological response of cells to a mitogenic preparation, it becomes essential to purify and adequately characterize the mitogenic proteins involved. If one is not working with a single, purified species, or at least a well-characterized combination of these, one cannot hope to faithfully reproduce or even accurately describe responses observed in culture. It is further apparent from these studies that crude plant extracts, though adequate for some mitogenic studies, must be viewed with considerable caution if a wide range of experiments or investigations from a variety of laboratories are to be closely correlated.

Lectins from Other Plant Species

I would now like to discuss some experiments done in collaboration with Dr. Stephen Kauffman at the Massachusetts Institute of Technology and Shirley Hoover at The Hershey Medical Center. About five years ago we became intrigued by the observation that among the most poisonous plants that abound in the tropics and subtropics a protein was alleged to be associated with the toxic principle. Our initial aim was to determine whether by chance these toxic proteins were also lectins, and we began by simply screening a variety of poisonous plant seed extracts for hemagglutinating ability. As it turned out, of the three plants on which we eventually focused our attention, all contained potent lectin-mitogens; but in no case were these strictly identical to the protein toxin. The three plants we studied were *Abrus precatorius* (rosary pea, crabs eye), *Hura crepitans* (sandbox tree) and *Robinia pseudoacacia* (black locust). Of the three, *Abrus precatorius* contains by far the most toxic protein, abrin, but also by far the most tenacious lectin, since it will agglutinate red blood cells from almost any species at μg/ml concentrations in minutes. Both *Robinia pseudoacacia* and *Hura crepitans* also possess lectins, but these show a pronounced effect only against blood cells that have been pretreated with trypsin to remove protective surface protein and expose receptor sites. In any case, after having established by agglutination inhibition tests that at least part of the sugar specificity of the three lectins in each case resided in a terminal galactose residue on the receptor glycopeptide, we isolated the lectin from each extract through the use of an appropriate chromatography resin.

Table 4.4   Stimulation of BALB/c Spleen Cells with Various Mitogens

Mitogen	Concentration (μg/ml)	CPM ^{3}H-Thymidine	Relative Stimulation
None	—	6,932	1.0
Abrin	0.03	105,947	15.3
Hurin	280	523,302	75.5
PHA	40	62,380	9.0
PWM	25	27,036	3.9

During the course of the isolations, we discovered that the purified lectin from *Abrus precatorius* as well as the crude extracts from both *Hura crepitans* and *Robinia pseudoacacia* were highly mitogenic in mixed lymphocyte cultures. In fact, using PHA and pokeweed mitogen (PWM) as standards, we found that the *Abrus* lectin gave levels of stimulation three times as high and crude *Hura* extract levels more than ten times as high as that of PHA and PWM (Kaufman and McPherson 1975).

To compare the efficiency of pure *Abrus* lectin and *Hura* extract as mitogens with PHA and PWM, BALB/c spleen cells were incubated with various concentrations of each, and the incorporation of ^{3}H-thymidine into DNA was measured. As is evident in table 4.4, the maximum stimulation attained with *Abrus* lectin and *Hura* extract was greater than that promoted by either PHA or by PWM. We consistently observed that *Hura* mitogen promoted between 60- and 150-fold stimulation and *Abrus* lectin promoted 12- to 17-fold increases in thymidine incorporation. PHA and PWM routinely yield 5–10- and 2.5–4.5-fold increases, respectively.

The responses of BALB/c spleen cells to various concentrations of *Abrus* and *Hura* mitogens (here called abrin and hurin, respectively, but see later discussion) are shown in figure 4.1. The maximum responses promoted by pure abrin was obtained at a final concentration of 30 ng/ml; at concentrations greater than 150 ng/ml, the preparation was toxic. Both *Abrus* lectin and *Hura* extract are active only within a narrow range of concentration, which is also true for other mitogens (Greaves and Janossy 1972; Raff 1971; Andersson, Sjöberg, and Möller 1972). At suboptimal concentrations of

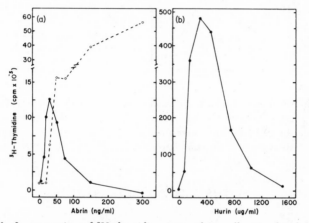

Figure 4.1. Incorporation of ³H-thymidine into spleen cells stimulated with Abrus lectin and Hura *lectin. Balb/c spleen cells are incubated with various concentrations of (a)* Abrus *lectin for 60 hr (●---●) or for 2 hr, then washed, and cultured an additional 58 hr (o---o) or with (b)* Hura *lectin for 60 hr (●---●).*

Abrus and *Hura* lectins, the kinetics of the responses were the same as those obtained with optimal doses.

To determine the nature of the specificity of the interaction between cell and mitogen, we sought to initially define the cell population that responded to both *Abrus* and *Hura* lectin. Spleen cells from athymic nude mice (nu/nu), which are severely depleted of functional populations of thymus-derived lymphocytes, were challenged with the mitogens. Heterozygous CBS (+/nu), litter mates were used as positive controls. As may be seen in figure 4.2, the mitogenic response to *Hura* and *Abrus* mitogen does require T cells, since spleens from nude mice, virtually devoid of T cells, show no appreciable stimulation over controls. These mitogens are, therefore, good probes for T cells.

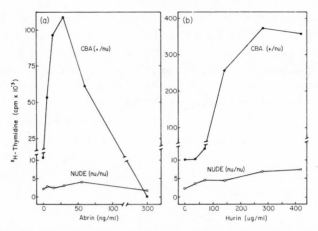

Figure 4.2. Stimulation of CBA and nude mice spleen cells with Abrus *and* Hura *lectin. Spleen cells from nude (nu/nu) mice or their heterozygous CBA (+/nu) litter mates were incubated with various concentrations of (a)* Abrus *lectin or (b)* Hura *lectin.*

Abrus Lectin and Abrin

The *Abrus* lectin was easily purified, since it bound directly to the galactose residues on unsubstituted Sepharose; elution with 0.2 M galactose yielded essentially pure lectin.

Co-purifying with the *Abrus* lectin was a portion of the protein toxin, known as abrin. At this point it is necessary to describe the probable relationship between the toxic protein abrin and the *Abrus* lectin, which has been shown to be strictly nontoxic (Olsnes and Pihl 1976). As stated earlier, the lectin and toxin co-purify using affinity techniques; thus, both bind galactose residues. Both are hemagglutinating proteins existing as dimers or tetramers, and these are composed of monomers of 64,000 daltons. In both cases the monomers consist of two polypeptide chains of weights 33,000 and 31,000. Both are glycoproteins, and the amino acid compositions are virtually if not completely identical (Olsnes and Pihl 1976). The toxin and the lectin do not seem to be separable on SDS polyacrylamide gels. The inescapable conclusion would seem to be that they are in fact one and the same, and in essence, this is true. One, however, is distinctly toxic, and the other is not. In addition, the lectin is strongly mitogenic. Thus, they are *not* identical.

In the case of abrin, the toxin protein monomer, it has been shown that the 33,000-dalton B chain is entirely responsible for the binding of the protein to the cell membrane, while only the 31,000-dalton polypeptide enters the cell and assumes, then, full responsibility for the toxic action (Olsnes and Pihl 1976). In the case of the lectin, the 33,000-dalton B chain, as one might expect, is again responsible for the binding of the protein to the cell membrane; the function of the 31,000-dalton A chain, however, is unknown in this case, even assuming that it has one.

Thus, two possibilities seem to exist: Either the fundamental protein is the highly toxic abrin and the *Abrus* lectin is derived from the toxin by some type of modification, or the situation is reversed and the toxin is derived from the lectin. Our evidence indicates that the former is true and that the lectin, whose mitogenic potential is inherent but superseded, is in fact derived by inactivation of the toxic 31,000 A chain of abrin.

We are, therefore, proposing that at some point a conversion from toxin to mitogen takes place that is virtually undetectable by most physical-chemical analyses. We further suggest that the nature of the modification may be either an aggregation of monomers into oligomeric structures or, more likely, changes in the poorly characterized polysaccharide moieties of the glycopeptides. Another alternative is that the A chains of the lectin and toxin differ by only a few amino acids owing to hydrolysis or mild proteolysis.

A very intriguing aspect of the abrin toxin is its similarity in structure, function, and apparent mechanism to the bacterial toxins, i.e., cholera, tetanus, and diphtheria (Olsnes and Pihl 1976). Each of these toxic proteins is also composed of two polypeptide chains linked by disulfide bonds, one of which is responsible for the binding of the protein to the cell surface and the other for the toxic activity. In common with abrin, all these bacterial toxins act at the ribosome level to disrupt and eventually destroy protein synthesis in the cell. Thus, it appears entirely possible that certain bacterial toxins and the class of plant toxins resembling abrin may have evolved from some common primordial precursor.

The contamination of the lectin by the toxin even at very low levels presented a serious problem in any mitogenic tests, however, since it was still at a sufficient concentration so that it had a profound destructive effect on the lymphocytes. After standing for several weeks at 4°C, however, the toxin apparently became degraded or modified and lost its effectiveness, while the lectin remained intact (Kaufman and McPherson 1975). This observation

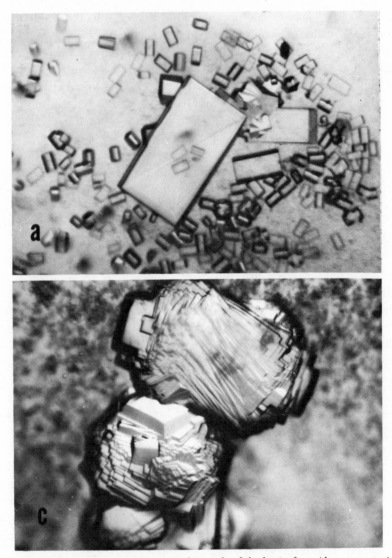

Figure 4.3. Single and multiple twinned crystals of the lectin from Abrus precatorius now under investigation by X-ray diffraction techniques.

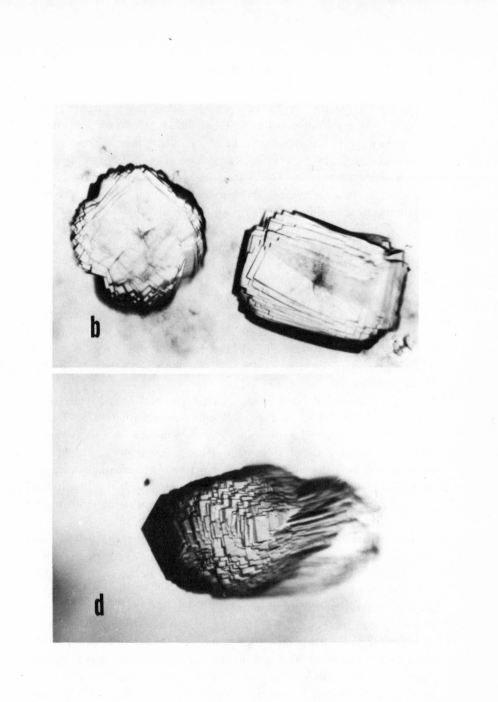

supports the contention that the toxic protein, abrin, is a subspecies of the lectin (i.e., it is the same polypeptide but differs either in a few amino acids or in its polysaccharide composition), and that it may be converted to the nontoxic lectin by proteolytic cleavage or loss of some sugars from its polysaccharide component. This is reinforced by the demonstration of Lin and associates that the protein abrin becomes nontoxic when its polysaccharide moiety is modified with periodate (Lin et al. 1970; Lin, Shaw, and Tung 1971; Lin et al. 1971).

The *Abrus* lectin is by far the lectin that is best structurally characterized (McPherson and Rich 1973; Olsnes and Pihl 1972, 1973; Olsnes, Heiberg, and Pihl 1973) as being composed of two polypeptide chains, a B chain of 33,000 daltons and an A chain of 31,000 daltons. While the B chain is essential for binding to the cell membrane, the A chain is homologous to that necessary for expression of the toxic effect in abrin. These two chains are joined by disulfide linkages to yield a protomer molecular weight of 64,000. In solution the protein exists as both a dimer and a tetramer, probably in equilibrium.

Crystals of the *Abrus* lectin-mitogen have been obtained and subjected to preliminary X-ray diffraction analysis (McPherson and Rich 1973). Examples of the crystals used for this study are shown in figure 4.3. A complete three-dimensional study of its structure to high resolution is currently in progress. The three-dimensional atomic structure of concanavalin A has already been determined by Edelman and colleagues (1972), and the lectins from the fava bean, the garden pea, and the castor bean are currently under study by X-ray diffraction techniques as well.

Microcrystals of the *Abrus* lectin have also been obtained, and this has permitted an electron microscopy study of the crystalline protein. Examples of negatively and positively stained crystals as seen in the electron microscope are shown in figures 4.4 and 4.5. These investigations indicate that the four identical subunits of the tetrameric molecule are grouped in a more or less square planar arrangement having at least one exact dyad axis with probable 222 molecular point group symmetry. The average diameter of the abrin molecule is 80 to 90 Å and the individual subunits appear to have a diameter of about 30 Å. If one assumes the receptor binding sites to be located toward the periphery of the molecule, then it is reasonable that the abrin tetramer could span adjacent sites on a single cell surface or on two separate cell membranes as distant as 100 Å.

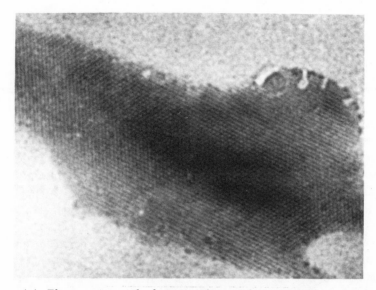

Figure 4.4. Electron micrograph of a negatively stained microcrystal of the Abrus lectin showing the individual protein monomers.

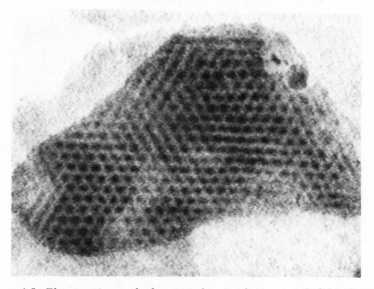

Figure 4.5. Electron micrograph of a positively stained microcrystal of the Abrus lectin showing the protein tetramers.

Lectins from *Hura* and *Robinia*

The lectins from *Hura crepitans* and *Robinia pseudoacacia* were considerably more difficult to isolate, as neither could be adsorbed onto the galactose residues of Sepharose directly. Success was achieved only after a large variety of affinity resins were devised and tested. Eventually, we found that *Hura* lectin could be reversibly bound to a column of cyanogen bromide activated Sepharose to which blood group substance from hog gastric intestinal mucosa was linked. Elution with saturated lactose (galactose alone was not effective) yielded a small quantity of pure protein. The recovery was about 10 mg of lectin from approximately 50 g of seeds (McPherson and Hoover, manuscript in preparation).

The lectin from *Robinia pseudocacia* was finally purified on a Sepharose column to which fetuin had been bound by cyanogen bromide treatment. In this case the terminal sialic acid residues of the fetuin had first to be removed by exposure to neuraminidase. Elution of this column with either galactose or lactose released a single pure protein. The recovery was once again about 10 mg of lectin from about 25 g of seeds (McPherson and Hoover, manuscript in preparation).

Although *Abrus* lectin was effective in agglutinating native erythrocytes from several species at very low concentrations, both *Hura* and *Robinia* lectins were effective only with erythrocytes that had been treated with trypsin. These results, along with the resin-binding characteristics, indicate that although the three lectins share a common affinity for galactose residues, *Hura* and *Robinia* lectins are more discriminating and probably bind a specific di- or oligosaccharide receptor. It might be noted in this regard that we also attempted to isolate *Hura* lectin and *Robinia* lectin by absorption onto galactosamine joined to Sepharose through a caproic acid stem and N-acetyl-galactosamine joined in the same fashion. In both of these trials, the lectins did not bind. *Robinia* lectin failed to adsorb to the blood group/Sepharose resin, while *Hura* lectin failed to bind to the fetuin/Sepharose column. This suggests that the specificities of the three lectins, although similar in the requirement for terminal galactose, are indeed significantly different. Neither *Hura* lectin nor *Robinia* lectin agglutinates even trypsinized erythrocytes at anywhere near the low level at which *Abrus* lectin proved effective. This might well reflect the greater number of available receptor sites on the cell's surface owing to *Abrus* lectin's less demanding binding specificity. Analysis of the three lectins for carbohydrate yielded a positive result in every case, indicating that all are glycoproteins. The amino acid composition of the three lectins has also been determined.

When run on adjacent tracks of a gradient slab gel in the presence of excess β-mercaptoethanol and SDS, *Abrus* lectin yields two bands that correspond to molecular weights of 34,000 and 31,000 daltons, as described previously. *Robinia* lectin shows a strikingly similar band pattern, yielding two polypeptide chains of 31,000 and 30,000 daltons. As shown by a second electrophoretic run in the absence of a reducing agent, these two polypeptides, unlike *Abrus* lectin, are not joined by disulfide linkages but are held together only by noncovalent bonds.

Hura lectin is different from both *Abrus* and *Robinia* lectins in that it exhibits only a single band on gel electrophoresis following reduction with β-mercaptoethanol. This single polypeptide, however, migrates almost exactly with the 33,000-molecular-weight B chain of the *Abrus* lectin. This correlation between the *Abrus* B chain and *Hura* lectin may be more than coincidence, since both are receptor binding chains, but *Hura* lectin, which is not toxic, is also not associated with an A chain, which in *Abrus precatorius* is implicated in the toxic principle.

Since both *Robinia* lectin and *Hura* lectin agglutinate, or cross-link, blood cells, they must be multivalent and therefore must be either dimers or tetramers. Thus, we can well expect that *Hura* lectin will exist as a dimer of about 70,000 daltons or a tetramer of 140,000 daltons and, similarly, that *Robinia* lectin will be a dimer of about 120,000 daltons or a tetramer of twice that molecular weight. The comparative polypeptide and subunit structures for the three mitogens are shown in figure 4.6.

We have only begun to characterize *Robinia* and *Hura* lectins in terms of their mitogenic potential, but our early results are somewhat disappointing. In preliminary experiments in which the dose response of murine spleen lymphocytes has not been optimized, we observed a stimulation in thymidine uptake about equal to that obtained with PHA using both purified *Robinia* lectin and purified *Hura* lectin. In the latter case, this is far less than that observed with the crude *Hura crepitans* seed extract. Although the level of stimulation may increase with modifications in the experimental conditions, it is not likely to do so by an order of magnitude. I would like to suggest that the reason for this may be either that the crude extract, like that from pokeweed, contains additional mitogenic proteins that act in concert with the lectin we isolated but that were not isolated by our procedure, or that the

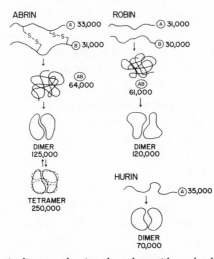

Figure 4.6. Schematic diagram showing the polypeptide and subunit organization in Abrus, Hura, *and* Robinia *lectins.*

crude extract contains proteases or other enzymes that modify the lympho-cyte surface membranes and make the cells more susceptible to binding by the lectins. The latter possibility seems attractive, since the *Hura crepitans* seed is known to contain several very active proteases and the agglutinating ability of *Hura* lectin, with red blood cells at least, is substantially enhanced by pretreatment with trypsin. Each of these possibilities will be examined in the near future.

REFERENCES

Andersson, J., G. Edelman, G. Möller, and O. Sjöberg. 1972. "Activation of B Lymphocytes by Locally Concentrated Concanavalin A." *Eur. J. Immunol.* 2:233–35.

Andersson, J., O. Sjöberg, and G. Möller. 1972. "Mitogens as Probes for Im-munocyte Activation and Cellular Cooperation." *Transplant Rev.* 11:131–77.

Basham, T. and M. Waxdal. 1975. "The Stimulation of Immunoglobulin Produc-tion in Murine Spleen Cells by the Pokeweed Mitogens." *J. Immunol.* 114:715–16.

Claman, H. 1970. "Cellular Immunology." In T. Baird, ed. *Immunology.* Kalama-zoo, Mich.: Upjohn.

Edelman, G., B. Cunningham, G. Reeke, J. Becker, M. Waxdal, and J. Wang.

1972. "The Covalent and Three-Dimensional Structure of Concanavalin A." *Proc. Nat. Acad. Sci. USA* 69:2580–84.

Edelman, G., I. Yahara, and J. Wang. 1973. "Receptor Mobility and Receptor–Cytoplasmic Interactions in Lymphocytes." *Proc. Nat. Acad. Sci. USA* 70:1442–46.

Fisher, D. and G. Mueller. 1971. "Studies on the Mechanism by Which Phytohemagglutinin Rapidly Stimulates Phospholipid Metabolism of Human Lymphocytes." *Biochem. Biophys. Acta* 248:434–48.

Greaves, M. and S. Bauminger. 1972. "Activation of T and B Lymphocytes by Insoluble Phytomitogens." *Nature (London) New Biol.* 235:67–70.

Greaves, M. and G. Janossy. 1972. "Elicitation of Selective T and B Lymphocyte Responses by Cell Surface Binding Ligands." *Transplant Rev.* 11:87–130.

Hersh, E. and J. Oppenheim. 1965. "Impaired *in Vitro* Lymphocyte Transformation in Hodgkin's Disease." *N. Engl. J. Med.* 273:1006–12.

Kaufman, S. and A. McPherson. 1975. "Abrin and Hurin: Two New Lymphocyte Mitogens." *Cell* 4:263–69.

Lin, J.-Y., S.-T. Ju, Y-S. Shaw, and T-C. Tung. 1970. "Distribution of I^{131} Labeled Abrin *in Vivo*." *Toxicon* 8:197–201.

Lin, J.-Y., Y-S. Shaw, and T-C. Tung. 1971. "Studies on the Active Principle from *Abrus precatorius* L. *Leguminosae* Seed Kernels." *Toxicon* 9:97–101.

Lin, J.-Y., Y-C. Cheng, K. Liu, and T-C. Tung. 1971. "Carbohydrate in Abrin." *Toxicon* 9:353–60.

McPherson, A. and A. Rich. 1973. "Studies on Crystalline Abrin: X-ray Diffraction Data, Molecular Weight, Carbohydrate Content and Subunit Structure." *FEBS Lett.* 35:257–61.

McPherson, A. and S. Hoover. "Purification of Mitogens from *Hura crepitans* and *Robinia pseudoacacia*." Manuscript in preparation.

Nowell, P. 1960. "Phytohemagglutinin: An Initiator of Mitosis in Cultures of Normal Human Leukocytes." *Cancer Res.* 20:462–66.

Olsnes, S. and A. Pihl. 1972. "Treatment of Abrin and Ricin with β-Mercaptoethanol. Opposite Effects of Their Toxicity in Mice and Their Ability to Inhibit Protein Synthesis in a Cell-Free System." *FEBS Lett.* 28:48–50.

Olsnes, S., R. Heiberg, and A. Pihl. 1973. "Inactivation of Eucaryotic Ribosomes by the Toxic Plant Proteins Abrin and Ricin." *Mol. Biol. Rep.* 1:15–20.

Olsnes, S. and A. Pihl. 1973. "Isolation and Properties of Abrin: A Toxic Protein Inhibiting Protein Synthesis (Evidence for Different Biological Functions of Its Two Constituent Peptide Chains)." *Eur. J. Biochem.* 35:179–85.

—— 1976. "Abrin, Ricin and Their Associated Agglutinins." In P. Cuatrecasas, ed., *Receptors and Recognition Series: The Specificity and Action of Animal, Bacterial and Plant Toxins*, pp. 129–74. London: Chapman and Hall.

Raff, M. 1971. "Surface Antigenic Markers for Distinguishing T and B Lymphocytes in Mice." *Transplant Rev.* 6:52–80.

Reisfield, R., J. Börjeson, L. Chessin, and P. Small. 1967. "Isolation and Characterization of a Mitogen from Pokeweed (*Phytolacca americana*)." *Proc. Nat. Acad. Sci. USA* 58:2020–27.

Sharon, N. and H. Lis. 1972. "Lectins: Cell-Agglutinating and Sugar-Specific Proteins." *Science* 177:949–59.

Siraganian, P. and R. Siraganian. 1974. "Basophil Activation by Concanavalin A: Characteristics of the Reaction." *J. Immunol.* 112:2117–25.

—— 1975. "Mechanism of Action of Concanavalin A on Human Basophils." *J. Immunol.* 114:886–93.

Taylor, R., W. Duffus, M. Raff, and S. de Petris. 1971. "Redistribution and Pinocytosis of Lymphocyte Surface Immunoglobulin Molecules Induced by Antiimmunoglobulin Antibody." *Nature (London) New Biol.* 233:225–29.

Waxdal, M. 1974. "Isolation, Characterization and Biological Activities of Five Mitogens from Pokeweed." *Biochemistry* 13:3671–77.

Waxdal, M. and T. Basham. 1974. "B and T-Cell Stimulatory Activities of Multiple Mitogens from Pokeweed." *Nature (London)* 251:163–64.

Waxdal, M., S. Nilsson, and T. Basham. 1976. "Heterogeneity of Pokeweed Mitogens and the Responding Lymphocytes." In J. Oppenheim and D. Rosenstreich, eds., *Mitogens in Immunobiology.* London: Academic Press.

5   Literature Review and Clinical Management of Household Ornamental Plants Potentially Toxic to Humans

Ara Der Marderosian and Frank C. Roia Jr.

Interest in poisonous plants has continued unabated in recent years, and in fact, for the first time in the United States they have assumed the highest category of substances most frequently ingested by children under the age of 5, as reported by the National Clearinghouse for Poison Control Centers (HEW, 1976). The category showing the largest percentage increase over the preceding year was insecticides. These rose almost 35 percent and climbed into eighth place on the top-25 list. Plants also rose a significant 20 percent over the year-earlier figure. According to Opp (1977), in 1975 there were 7,710 ingestions of plants reported to the National Clearinghouse. Of this group, 905 persons had symptoms, 186 were admitted to hospitals, and 3 died. Of course, many cases are not reported, so the figures are probably much higher. Reasons given for the increase include greater interest in gardening, floriculture, and simple pursuits of nature such as hiking, climbing, camping, and eating wild plants.

It is interesting to note that relatively few phytotoxicological studies have focused specifically on house plants, perhaps the most readily available group of plants to humans. For this reason we have attempted to carry on

Acknowledgement is made to the valuable assistance lent by Dr. Julia F. Morton, University of Miami, and Dr. John M. Kingsbury in procuring many of the references used in this article.

This investigation was supported by a grant from the Longwood Foundation, Kennet Square, Pennsylvania.

For the previous article in this series, see A. Der Marderosian, F. Giller, and F. C. Roia, Jr. 1976. "Phytochemical and Toxicological Screening of Household Plants Potentially Toxic to Humans. 1." *J. Toxicol. Environ. Health* 1:939–53.

positive studies to reveal whether or not certain of these carry any potential hazard. Notwithstanding the classical problems in phytotoxicology, as outlined by Kingsbury (1969), we have attempted to develop toxicological profiles of house plants so that the development of clinical management of their ingestion can proceed on a rational basis.

Direct Investigation of the Toxicity of Household Plants

In a previous investigation thirty different genera of household ornamental plants were studied from a phytochemical and toxicological point of view (Der Marderosian, Giller, and Roia 1976). Different plant parts were tested for alkaloids, glycosides, steroids, saponins, and cyanogenic glycosides. At least seven genera showed detectable concentrations of alkaloids. The glycoside tests were variable, and there was an indication that several species contained steroidal compounds. Sixteen genera showed positive saponin foam tests. None of the species was clearly cyanogenic. Preliminary biological screening data on their toxicity to rats and mice were determined. Of the plants studied (see table 5.1), those that have a well-known history of being poisonous were confirmed as such. These included species of *Caladium*, *Convallaria*, *Dieffenbachia*, *Digitalis*, *Hydrangea*, *Lantana*, *Nerium*, *Philodendron*, *Podophylum*, and *Solanum*. Others not previously reported to be poisonous (e.g., *Begonia* hyb. C. P. Raffill, leaf and stem; *Codiaeum variegatum* Muell., var. *pictum* cv. Mr. Peters, leaf; *Dracaena sanderiana* Hort. Sand., leaf; *Ficus lyrata* (*pandurata*) Hort., leaf; *Gladiolus gandavensia* Van Houtte, flowers; *Hippeastrum* hybrid, flowers; *Maranta leuconeura* Morr. var. *kerchoveana*, leaf; *Monstera deliciosa* Liebm., leaf; *Peperomia obtusifolia* Dietr. cv. Aurea, leaf; *Pilea cadierei* Gagnep. and Guillaumin, leaf and stem; *Sansevieria trifaciata* Praim. var. *laurentii*, leaf and flower; *Sansevieria thyrsiflora* Thunb., leaf; *Selaginella pallescens* Presl., leaf; and *Syngonium podophyllum* Schott. var. *albo-virens*, leaf and stem) may apparently be toxic. However, it should be pointed out that relatively large quantities (3 g/100 g body weight) were administered to the rats intraperitoneally (i.p.) in order to determine *any* potential for toxicity that may exist for these plants. They may therefore not be as poisonous as observed here, because i.p. administration generally produces much greater absorption of toxic principles than oral administration. In fact, close examination of the examples in table 5.1 shows that while i.p. toxicity in rats was confirmed in some cases by oral toxicity in mice, there were also several instances where this was not so. As stated previously (Der Marderosian, Giller, and Roia 1976), closer scrutiny is needed for the plants that show toxic potential in this toxicological screening study. What is specifically needed are LD_{50} determinations by an oral route in several animal species before meaningful extrapolations can be made to possible injurious effects in man. At this point those that show no toxicity in the toxicological screen in both rats and mice are *Ficus elastica* (*decora*) Roxbg., leaf and stem, and *Nephrolepis exaltata* Scott cv. Scottii, leaf. Those that showed no toxicity in the mice oral route or rat i.p. route alone include *Anthurium andraeanum* Lind., leaf; *Arisaema triphyllum* (L.) Schott, leaf and petiole; *Begonia* hyb. cv. Bow arriola, leaf and petiole combined; *Euphorbia pulcherrima* Willd., flowers, leaves, and bracts; *Euphorbia tirucalli* L., stem; *Hemerocallis lilio-asphodelus*, flower buds, flowers, leaf; *Hydrangea arborescens* L., flowers; *Hippeastrum* hybrid, leaf; *Paeonia lactiflora* Pall., stem and leaf combined and flowers; *Philodendron sagittifolium* Liebm., leaf and immature fruit; *Saintpaulia ionantha* Wendl. cv. Ocean Waves, leaf and flowers combined; and *Xanthosoma hoffmannii* Schott, leaf. Of those listed, *Arisaema triphyllum* (jack-in-the-pulpit), *Euphorbia pulcherrima* (poinsettia), *Euphorbia tirucalli* (pencil bush) and *Hippeastrum* species are well known to contain locally irritating substances that have harmed humans when sufficient quantities were ingested or handled improperly. So it is difficult to extrapolate directly from the limited toxicological data of this preliminary mouse and rat screen. On the other hand, *Ficus elastica* (*decora*) (rubber plant), *Nephrolepis exaltata* (Boston fern), *Anthurium andraeanum*, *Hemerocallis lilio-asphodelus* (day lily), *Paeonia lactiflora*, *Saintpaulia ionantha* (African violet), and *Xanthosoma hoffmannii* are not known to be harmful to humans generally, and the preliminary data at least bear this out. Again, it is emphasized that follow-up studies need to be conducted to complete the toxicity profile of these plants. It should, finally, be mentioned here, with respect to the plants listed in table 5.1, that several non-house plants were included both because they were available at the time of the study and because they served as either "positive" or "negative" controls. Blank areas in the table indicate that insufficient material was available at the time of the experiments.

Table 5.1 Preliminary Results of Acute Toxicity Studies of House Plant Extracts on Rats and Mice (Der Marderosian, Giller, and Roia 1976)

Plant	Longwood Accession no.	Plant Part[a]	Rats i.p.[b]	Mice Oral[c]
1. *Anthurium andraeanum* Lind.	—	L		0/3[d]
2. *Arisaema triphyllum* (L.) Schott	—	L		0/3
		P		0/3
3. *Begonia* hyb. cv. Bow arriola	561660	L & P	0/3	
4. *Begonia* hyb. C. P. Raffill	581326	L	3/3	
		S	3/3	
5. *Caladium bicolor* (Ait.) Vent.	—	L	4/4	
		S	0/3	
6. *Codiaeum variegatum* Muell. var.	611643	L	0/3	
pictum cv. Mr. Peters		L (im)	3/3	
7. *Convallaria majalis* L.	—	Fr (U)	3/3	
8. *Dieffenbachia amoena* Gentil	562136	L	2/3	0/3
		S	3/3	
9. *Dieffenbachia bausei* Regel	—	L		3/6
10. *Dieffenbachia picta* Schott	55473	L	3/3	2/6
		S	3/3	
11. *Dieffenbachia picta* Schott	55490	L	3/3	
cv. Rudolph Roehrs		S	3/3	
12. *Digitalis purpurea* L.	—	L		2/5
		F		0/3
13. *Dracaena sanderiana* Hort. Sand.	55489	L	4/4	
14. *Euphorbia pulcherrima* Willd.	—	F		0/3
		Br		0/3
		L		1/6
15. *Euphorbia tirucalli* L.	—	S	0/4	
16. *Ficus elastica* (*decora*) Roxbg.	562138	L	0/3	0/6
		S		0/3
17. *Ficus lyrata* (*pandurata*) Hort.	5520	L	7/7	
		Fr (U)	0/2	
18. *Gladiolus gandavensia* Van Houtte	—	F	3/3	
19. *Hemerocallis lilio-asphodelus*	—	F (buds)		0/3
(*flava*) L.		F		0/3
		L		0/3
		R		2/6
20. *Hippeastrum* hybrid	—	L		0/3
		F	6/6	
21. *Hydrangea arborescens* L.	—	F		0/3
22. *Hydrangea macrophylla* Ser. (blue	—	F	4/4	0/3
flowers)		L	3/3	
23. *Lantana camara* L. cv. Gold mound	59569	Fr (U)	4/5	
		Fr (R)	2/2	
		F		0/3

Table 5.1 continued

Plant	Longwood Accession no.	Plant Part[a]	Rats i.p.[b]	Mice Oral[c]
24. *Maranta leuconeura* Morr. var. *kerchoveana*	55494	L	4/4	
25. *Monstera deliciosa* Liebm.	L.509	L	4/5	2/6
26. *Nephrolepis exaltata* Scott cv. Scottii	L.2758	L	0/3	0/3
27. *Nerium oleander* L. (red flower)	L.1827	L		10/10
		F	2/2	
28. *Nerium oleander* L. (double pink flower)	L.79	F	3/3	
		L	3/3	
29. *Nerium oleander* L. (white flower)	L.1828	F	2/2	
30. *Paeonia lactiflora* Pall.	—	S & L		0/3
		F		0/3
31. *Peperomia obtusifolia* Dietr. cv. Aurea	55476	S & L		0/3
		L	5/6	3/6
32. *Philodendron oxycardium* Schott	55526	L	5/6	3/6
		S	2/8	0/3
33. *Philodendron sagittifolium* Liebm.	—	L		0/3
		F		0/3
34. *Pilea cadierei* Gagnep. & Guillaumin	5781	L	3/6	
		S	3/6	
35. *Podophylum peltatum* L.	—	Fr (U)	2/3	
		Fr (R)	2/2	
36. *Saintpaulia ionantha* Wendl. cv. Ocean Waves	591833	L	0/3	
		L & F	0/3	
37. *Sansevieria trifaciata* Praim. var. *laurentii*	56269	L	3/3	0/3
		F	4/4	2/6
38. *Sansevieria thyrsiflora* Thunb.	—	L		3/6
39. *Selaginella pallescens* Presl.	5510	L	5/6	
40. *Solanum pseudocapsicum* L.	—	L		0/3
		Fr (R)	3/5	0/3
		Fr (U)	4/5	
		W		1/5
41. *Syngonium podophyllum* Schott. var. *albo-virens*	55483	L	8/8	1/3
		S	7/9	
42. *Xanthosoma hoffmannii* Schott.	592685	L	0/3	

[a]Br, bract; C, corm; F, flowers; F (im), flowers (immature); Fr (R), fruits (ripe); Fr (U), fruits (unripe); L, leaves; L (im), leaves (immature); P, petioles; R, roots, S, stem; W, whole plant in flower excluding the roots.

[b]Deaths following i.p. administration of 3.0 g plant extract (contained in 10.0 ml liquid suspension/100 g body weight). Blank space indicates plant or plant part not available for testing.

[c]Deaths following oral administration of 100 mg lyophilized plant material (suspended in sufficient distilled water to allow passage through oral feeding needle)/35 g body weight.

[d]Fractional notations; numerator = no. of animals dead; demoninator = no. of animals tested.

Toxicological Information on Household Plants

Another way to develop a toxicological profile on plants is to gather all the readily available literature and assemble it in tabular form for comparison as to active principles and toxicological experiences. This has been done for the household plants looked at in this investigation. Table 5.2 provides the Latin name for the plants, the active principles reported, toxicity or edibility data, and the literature reference found for these. In cases in which specific species names could not be located, related species were listed to get some idea of how these may have been used in their countries of origin. Many of our household plants are of tropical or subtropical origin, and several of the references were texts that listed economic uses of these plants. Dashes in the table simply indicate that no data were given. Table 5.2 generally reveals that little is known about many of the common household plants, and in fact, few or no references could be found at all for some species (e.g., *Anthurium andraeanum, Maranta leuconeura, Nephrolepis exaltata, Paeonia lactiflora, Peperomia obtusifolia, Pilea cadierei, Saintpaulia ionantha, Selaginella pallescens, Syngonium podophyllum,* and *Xanthosoma hoffmannii*). Obviously, more study is needed of these, since they are all fairly common house plants.

On the other hand, as expected, numerous references were found for *Arisaema, Begonia, Caladium, Codiaeum, Convallaria, Dieffenbachia, Digitalis, Dracaena, Euphorbia, Ficus, Gladiolus, Hemerocallis, Hippeastrum, Hydrangea, Lantana, Monstera, Nerium, Philodendron, Podophyllum, Sansevieria,* and *Solanum* species. The data for these generally overlapped and confirmed either toxicity or lack of it. In addition, many have either well known or rather obscure medicinal uses in the countries of origin. These data may also be of value in toxicological assessment, since they often represent the only data available on human experience with the plants.

Among the generalizations that can be made using information gathered in this way is the fact that *Anthurium* and *Begonia* species appear to be relatively nontoxic. No reports of direct, serious toxic experiences were observed. Little is known about any active principles. Some of the *Begonia* species have been used as diuretics, purgatives, or emetics. Several species are reported to be edible, particularly as young, fresh leaves or boiled as vegetables. Much folkloric information is available on *Arisaema triphyllum* (jack-in-the-pulpit) with respect to its orally irritating qualities, which it shares with other members of the Araceae. However, as indicated in table 5.1, the leaf and petiole were nontoxic in mice.

Interestingly, while *Caladium* species are generally acknowledged to be toxic owing to their calcium oxalate and/or a toxic protein constituent, several references revealed that these plants are eaten as a vegetable in tropical America and the West Indies when cooked.

Morton (1962) has also reported that *Codiaeum* species (crotons) are edible and that varietal differences in the acridity of the leaves exist. Yellow varieties are reported to be sweet and edible when young in the East Indies. Normally, however, the horticultural varieties available as house plants apparently contain irritant juices that may be noxious or allergenic. Medicinal uses include purgative, abortifacient, antitussive, and other uses, and a sudatory effect from a decoction.

Convallaria majalis L. (lily-of-the-valley) is widely acknowledged through several studies over the years to be toxic owing to its content of cardioactive glycosides convallarin and convallamarin. Table 5.2 reveals numerous reports of its hazardousness as a house or garden plant. So there is little doubt that all parts of this plant should be treated cautiously.

Similarly, almost all species of *Dieffenbachia* have well-recorded histories of causing concern, particularly when the stem is bitten or eaten. The matter of which species or varieties may be more or less hazardous has not yet been resolved. A number of recent studies have focused on this genus. Walter and Khanna (1972) have reported the presence of a proteolytic enzyme in three species of *Dieffenbachia* (*D. seguine*, *D. amoena*, and *D. picta*) and have named it "dumbcain." They suggest that the proteolytic action, with subsequent release of kinins, may be partially or wholly responsible for the activity of the plant. The kinins can act on reproductive ducts and may help explain the activity of *Dieffenbachia* species in producing sterility in human subjects as well as in mice and rabbits. Proteolytic enzymes are also known to cause severe local necrosis and destruction of small blood vessels, producing hemorrhage. Walter and Khanna believe that when these plants are swallowed or rubbed, sharp, pointed raphides of calcium oxalate cause the initial injury, whereupon the proteolytic enzyme ("dumbcain") enters the injured tissues to produce itching, pain, and swelling. In the mouth this results in swelling, salivation, and local necrosis, ultimately causing loss of ability to speak. They also suggest that at the injured site the raphides themselves may activate kinin-releasing mechanisms in the body. In another study Fochtman and associates (1969) studied the toxicity of *Dief-*

Table 5.2 Literature Review of Household Ornamental Plants and Related Species, Toxicity/Edibility Data

Plant	Principles Reported	Toxicity/Edibility Data	Literature Reference
1. *Anthurium andraeanum* Lind.	—	Nontoxic (see table 5.1)	Der Marderosian, Giller, and Roia 1976
2. *Arisaema triphyllum* (L.) Schott	—	Nontoxic (see table 5.1)	*Ibid.*
3. Same	Calcium oxalate	Roots cause severe burning to throat; plant causes dermatitis	Kingsbury 1964; Hardin and Arena 1974
4. *Begonia barbata* Wall	—	Leaves and stems edible	Sturtevant 1919
5. *Begonia cucullata* Willd.	—	Leaves edible	*Ibid.*
6. Same	—	Diuretic (plant)	Lewis and Elvin-Lewis 1977
7. *Begonia gracilis*	—	Irritant, emetic, purgative, diuretic	Lewin 1962
8. *Begonia hirtella* Link.	—	Leaves edible	Burkill 1935
9. *Begonia* hyb. cv. Bow arriola	Alkaloid	Nontoxic (see table 5.1)	Der Marderosian, Giller, and Roia 1976
10. *Begonia* hyb. C.P. Raffill	Alkaloid	Toxic (see table 5.1)	*Ibid.*
11. *Begonia malabarica* Lam.	—	Leaves edible	Sturtevant 1919
12. *Begonia picta* Sm.	—	Leaves edible	*Ibid.*
13. *Begonia rex*	—	Used as substitute for rhubarb, juice toxic to leeches	Chopra, Badhwar, and Ghosh 1965
14. Same	—	Irritant, emetic, purgative, diuretic	Lewin 1962
15. *Begonia sanguinea*	—	Diuretic (plant)	Lewis and Elvin-Lewis 1977
16. *Begonia* spp.	—	Two infant ingestions, no data	O'Leary 1964
17. *Begonia* spp. (several)	—	Leaves edible	Anonymous 1948
18. *Begonia sutherlandii* Hook.	Oxalic acid (?)	Emetic, purgative	Watt and Breyer-Brandwijk 1962
19. *Begonia tuberosa* Lam.	—	Leaves edible	Burkill 1935
20. *Caladium bicolor* (Ait.) Vent.	—	Toxic (see table 5.1)	Der Marderosian, Giller, and Roia 1976

21. Same	—	Leaves and corms edible	Sturtevant 1919
22. Same	Corrosive fluid	Enteritis	Lewin 1962
23. Same	—	Leaves edible	Terra 1966
24. Same	Calcium oxalate	Leaf and bulb edible cooked but not raw	Morton 1962
25. *Caladium colocasia*	—	Raw plant burns mouth but edible cooked	Arnold 1944
26. *Caladium picturatum* Koch	See *Caladium bicolor* (24) for details		Morton 1962
27. *Caladium* spp.	Irritant juice, calcium oxalate	Severe discomfort upon chewing	Kingsbury 1964; Hardin and Arena 1974; Lewis and Elvin-Lewis 1977
28. Same	—	Ingested by two children; nausea, vomiting	O'Leary 1964
29. *Codiaeum* spp.	Caustic latex	Contact dermatitis (?)	Lewis and Elvin-Lewis 1977
30. *Codiaeum variegatum* Bl.	6–8% tannin in latex	Juice used as purgative, abortifacient; decoction used as sudative, antitussive; very young leaves of certain yellow races eaten as flavoring	Burkill 1935
31. Same	—	Top shoots edible	Terra 1966
32. Same	—	Bark and roots cause burning in mouth if chewed; young leaves of certain yellow varieties edible; flower chewed by three children, causing slight irritation to mouth; whole leaf swallowed with no ill effects after intact retrieval with dose of ipecac	Morton 1962
33. Same	—	Bark and leaves cyanophoric	Quisumbing 1951

Table 5.2 continued

Plant	Principles Reported	Toxicity/Edibility Data	Literature Reference
34. Same	—	Used medicinally but considered to have poisonous properties	Irvine 1961
35. Same	—	Proximity to or handling produces active allergic dermatitis	Allen 1943
36. *Codiaeum variegatum* Muell. var. *pictum* cv. Mr. Peters	Alkaloid (?)	Toxic (see table 5.1)	Der Marderosian, Giller, and Roia 1976
37. *Convallaria majalis* L.	—	Toxic (see table 5.1)	*Ibid.*
38. Same	Cardioactive glycosides convallatoxin, convalloside	Toxic leaves, flowers, roots, and fruits; used medicinally as cardiotonic; emetic and purgative in high doses	Kingsbury 1964; Hardin and Arena 1974; Lewis and Elvin-Lewis 1977; Watt and Breyer-Brandwijk 1962
39. Same	—	Seeds, berry, pod ingested by four children; no data	O'Leary 1964
40. Same	Convallarin and convallamarin	"Five-year-old child once died from drinking water in which a bunch of lilies-of-the-valley had been standing"	Schenk 1955
41. Same	Convallamarin, convallarin, convallatoxin	LD or MLD or these glycosides for different animals	Spector 1955
42. *Dieffenbachia amoena* Gentil	—	Toxic (see table 5.1)	Der Marderosian, Giller, and Roia 1976
43. Same	Calcium oxalate and proteolytic enzyme complex	Proteolytic enzyme complex initiates itching, swelling, pain, and salivation via release of kinins; calcium	Walter and Khanna 1972

	Toxic constituent	Symptoms	Reference
44. *Dieffenbachia bausei* Regel	—	oxalate raphides facilitate rupture of tissue and penetration of toxic enzymes Toxic (see table 5.1)	Der Marderosian, Giller, and Roia 1976
45. *Dieffenbachia picta* Schott	Calcium oxalate, toxic protein, irritant juice	Toxic (see table 5.1)	Ibid.
46. Same	Calcium oxalate	Severe discomfort if chewed; stems more toxic than leaves; irritant dermatitis	Kingsbury 1964; Hardin and Arena 1974; Lewis and Elvin-Lewis 1977
47. Same	Calcium oxalate	Two ingestions ages 1 year and 18 years; temporary paralysis of throat muscles and burning due to calcium oxalate	O'Leary 1964
48. Same	See *Dieffenbachia amoena* (43) for details	—	Walter and Khanna 1972
49. Same	Calcium oxalate	Toxic to guinea pigs	Ladeira, Andrade, and Sawaya 1975
50. *Dieffenbachia picta* Schott cv. Rudolph Roehrs	—	Toxic (see table 5.1)	Der Marderosian, Giller, and Roia 1976
51. *Dieffenbachia seguine* (Jacq.) Schott	See *Dieffenbachia picta* (46) for details	—	Kingsbury 1964; Hardin and Arena 1974; Lewis and Elvin-Lewis 1977
52. Same	See *Dieffenbachia amoena* (43) for details	—	Walter and Khanna 1972
53. *Digitalis purpurea* L.	—	Toxic (see table 5.1)	Der Marderosian, Giller, and Roia 1976

Table 5.2 continued

Plant	Principles Reported	Toxicity/Edibility Data	Literature Reference
54. Same	Cardiac glycosides digitoxin, etc.	Human toxicity, usually from drug overdose; rarely children ingest flowers, seeds, or leaves; symptoms include nausea, diarrhea, abdominal pain, gross disturbances in heartbeat and pulse, mental irregularities, or tremors and convulsions	Kingsbury 1964; Hardin and Arena 1974; Lewis and Elvin-Lewis 1977; Watt and Breyer-Brandwijk 1962
55. *Dracaena cinnabari*	Resin	Resin used to stop hemorrhages	Lewis and Elvin-Lewis 1977
56. *Dracaena cylindrica* Hook. f.	—	Infusion of leaves poisonous, causing ulceration of stomach	Irvine 1961
57. *Dracaena fragrans* Gawl	Saponins	Root chewed; juice swallowed for postparturient pain	Watt and Breyer-Brandwijk 1962
58. *Dracaena laxissima* Engl.	—	Leaves used medicinally; sap said to blind people	Dalziel 1948
59. *Dracaena sanderiana* Hort. Sand.	Alkaloids (?)	Toxic (see table 5.1)	Der Marderosian, Giller, and Roia 1976
60. *Dracaena steudtneri* Engl.	—	Root used for rheumatism; leaf used for flatulence; no digitalis cardiac action	Watt and Breyer-Brandwijk 1962
61. *Euphorbia lactea*	Euphorbon, caoutchouc, taraxasterol, tirucallol, resin, gum euphorbium, rubber	Sap experimentally produced severe impairment of visual acuity, producing keratoconjunctivitis and uveitis	Crowder and Sexton 1964

62. *Euphorbia pulcherrima* Willd.	—	Nontoxic (see table 5.1)	Der Marderosian, Giller, and Roia 1976
63. Same	—	Ingestion of one poinsettia leaf resulted in death of child	Kingsbury 1964
64. Same	—	Many reports of ingestion of fruits, buds, and leaves received by poison control centers, but occasional local irritation (abdominal pain with vomiting) reported	Hardin and Arena 1974
65. Same	Alkaloids (?) (leaf); resins, caoutchouc (latex)	Latex highly irritant; used as depilatory; bark, leaf, and root regarded as markedly toxic	Watt and Breyer-Brandwijk 1962
66. Same	—	Leaf ingestion fatal to child; pods and leaf ingestion by two other children; no data	O'Leary 1964
67. Same	Germanicol, β-amyrin, pseudo-taraxasterol, pulcherrol, octaeicosanol, β-sitosterol, anthocyanin	Historic galactopoietic and abortive uses	Dominquez et al. 1967
68. Same	—	Large doses of homogenates from leaves, bracts, or flowers nontoxic to rats	Stone and Collins 1971
69. Same (and related species)	—	Toxic principles cause severe poisoning if ingested in quantity	Lewis and Elvin-Lewis 1977
70. *Euphorbia tirucalli* L.	—	Nontoxic (see table 5.1)	Der Marderosian, Giller, and Roia 1976

Table 5.2 continued

Plant	Principles Reported	Toxicity/Edibility Data	Literature Reference
71. Same	—	Juice irritant, but no cases of severe poisoning on record	Kingsbury 1964
72. Same	Caoutchouc, resin, taraxasterol (in latex)	Temporary blindness from latex; used for impotency, mosquito repellent, insecticide, fish poison, emetic, antisyphilitic; one adult death from hemorrhagic gastroenteritis	Watt and Breyer-Brandwijk 1962
73. Same	See *Euphorbia lactea* (61) for details		Crowder and Sexton 1964
74. Same	Rubber (in latex)	Latex irritant to skin and eyes (causing temporary blindness); poisonous internally	Burkill 1935
75. Same	Euphorbon	Latex irritant to skin and eyes (causing temporary blindness); poisonous internally	Morton 1958
76. Same (and related species)	Complex esters related to diterpene phorbol; may be cocarcinogens	Toxic principle in milky sap cause of dermatitis and severe poisoning if eaten in quantity; severe irritation to mouth, throat, and stomach by several species	Hardin and Arena 1974; Lewis and Elvin-Lewis 1977
77. *Ficus barclayana*	—	Fruit used for toothache	Lewis and Elvin-Lewis 1977
78. *Ficus carica*	Proteolytic enzyme ficin	Known or suspected to cause dermatitis and photodermatitis; latex used as anthelmintic	Hardin and Arena 1974; Lewis and Elvin-Lewis 1977
79. *Ficus cotinifolia*	Milky juice	Latex and bark used for wounds	Lewis and Elvin-Lewis 1977
80. *Ficus elastica (decora)* Roxbg.	—	Nontoxic (see table 5.1)	Der Marderosian, Giller, and Roia 1976

81. Same	—	Fig ingested by child; no data	O'Leary 1964
82. *Ficus glabrata*	Proteolytic enzyme ficin	Enzyme in latex used as anthelmintic	Lewis and Elvin-Lewis 1977
83. *Ficus hauili*	—	Irritant sap	Kalaw and Sacay 1925
84. *Ficus laurifolia*	See *Ficus glabrata* (82) for details	—	Lewis and Elvin-Lewis 1977
85. *Ficus lyrata (pandurata)* Hort.	—	Toxic (see table 5.1)	Der Marderosian, Giller, and Roia 1976
86. *Ficus platyphylla*	—	Dried latex chewed	Lewis and Elvin-Lewis 1977
87. *Ficus retusa*	—	Roots used for toothache	*Ibid.*
88. *Gladiolus ecklonii* Lehm.	—	Used for rheumatic pain	Watt and Breyer-Brandwijk 1962
89. *Gladiolus edulis* Burch.	—	Corm edible	*Ibid.*
90. *Gladiolus gandavensis* Van Houtte	—	Toxic (see table 5.1)	Der Marderosian, Giller, and Roia 1976
91. *Gladiolus ludwigii* Pappe	—	Decoction of corm used as enema to relieve dysmenorrhoea	Watt and Breyer-Brandwijk 1962
92. *Gladiolus multiflorus* Bak.	—	Decoction of corm used for dysentery	*Ibid.*
93. *Gladiolus psittacinus* Hook.	—	Decoction of corm used for colds and dysentery	*Ibid.*
94. *Gladiolus saundersii* Hook. f	—	Cooked corm used for diarrhea	*Ibid.*
95. *Gladiolus* spp.	—	Reported toxic in old reference	Kingsbury 1964
96. Same	—	Corm ingestion caused vomiting; leaf ingestion, no data	O'Leary 1964
97. *Hemerocallis lilio-asphodelus (flava)* L.	—	Nontoxic (see table 5.1)	Der Marderosian, Giller, and Roia 1976
98. *Hemerocallis* spp.	—	Flowers edible; no data on leaves or tubers; one ingestion, no data	O'Leary 1964
99. *Hippeastrum* hybrid	—	Toxic (see table 5.1)	Der Marderosian, Giller, and Roia 1976

Table 5.2 continued

Plant	Principles Reported	Toxicity/Edibility Data	Literature Reference
100. *Hippeastrum equestre*	—	Bulbs known to kill humans and animals in 2 to 3 hours	Lewis and Elvin-Lewis 1977
101. *Hippeastrum* spp.	Active alkaloids in family	Bulbs in this family known to cause gastrointestinal upset with possible trembling, convulsions, and death	Kingsbury 1964; Hardin and Arena 1974; Watt and Breyer-Brandwijk 1962
102. *Hippeastrum vittata* Herb.	Alkaloids, hippeastrine, lycorine, etc.	Bulb ingestion by three children; one vomited and fell asleep, but prompt stomach lavage avoided further effects	Morton 1962
103. *Hydrangea arborescens* L.	Glycosides (?)	Nontoxic (see table 5.1)	Der Marderosian, Giller, and Roia 1976
104. *Hydrangea macrophylla* Ser.	Glycosides (?)	Toxic (see table 5.1)	Ibid.
105. Same	—	Leaf ingestion caused illness	O'Leary 1964
106. *Hydrangea* spp.	Cyanogenic glycosides (?), hydrangin in leaves and buds; quinazolone alkaloid	Nausea and painful gastroenteritis and diarrhea in humans and animals; roots once used in dyspepsia	Kingsbury 1964; Hardin and Arena 1974; Lewis and Elvin-Lewis 1977; Watt and Breyer-Brandwijk 1962
107. *Lantana camara* L. cv. Gold mound		Toxic (see table 5.1)	Der Marderosian, Giller, and Roia 1976
108. *Lantana camara* L.	Lantadene A and B	Hepatogenic photosensitizer with gastroenteritis in animals; unripe fruit most dangerous; other symptoms are weakness, circulatory collapse, and death; acute symptoms resemble atropine poisoning	Kingsbury 1964; Hardin and Arena 1974; Lewis and Elvin-Lewis 1977; Watt and Breyer-Brandwijk 1962

#	Species	Toxic compound	Description	Reference
109.	Same	—	Three ingestions of berry and one of plant by children; no data	OLeary 1964
110.	Same	Lantadene A and B	Highly toxic to grazing animals, causing photosensitization; ripe fruits eaten by humans; unripe fruits toxic to children; leaf decoction used for rheumatism, colds, and indigestion	Morton 1962
111.	Same	Same	A number of ingestions of green berries resulted in acute poisoning and one death; resembles belladonna poisoning; treatment given	Wolfson and Solomons 1964; Wolfson 1964
112.	*Maranta leuconeura* Morr. var. *kerchoveana*	—	Toxic (see table 5.1)	Der Marderosian, Giller, and Roia 1976
113.	*Monstera deliciosa* Liebm.		Toxic (see table 5.1)	*Ibid.*
114.	Same	Calcium oxalate, irritant juice	Leaves and stems dangerous if eaten; all cultivated aroids may cause severe swelling of throat and mouth	Hardin and Arena 1974; Lewis and Elvin-Lewis 1977
115.	Same	—	Consumption of ripe fruit may cause allergy or anaphylaxis; extensive urticaria possible	Webb 1948
116.	Same	Calcium oxalate	Fruit edible when perfectly mature; unripe fruit may cause very unpleasant irritation of throat	Dahlgren 1947
117.	*Nephrolepis exaltata* Scott cv. Scottii	—	Nontoxic (see table 5.1)	Der Marderosian, Giller, and Roia 1976
118.	*Nerium oleander* L.	Glycosides (?)	Toxic (see table 5.1)	*Ibid.*
119.	Same	Glycosides oleandroside, nerioside	Toxic in all parts, green or dry, to animals and humans; single leaf considered potentially lethal to humans; nectar and honey poisonous; burning plant toxic; dermatitis possible	Kingsbury 1964; Hardin and Arena 1974; Lewis and Elvin-Lewis 1977; Watt and Breyer-Brandwijk 1962

Table 5.2 continued

Plant	Principles Reported	Toxicity/Edibility Data	Literature Reference
120. Same	—	Several cases of leaf ingestions; no data	O'Leary 1964
121. *Paeonia lactiflora* Pall.	—	Nontoxic (see table 5.1)	Der Marderosian, Giller, and Roia 1976
122. *Paeonia* spp.	—	Reported toxic in old reference	Kingsbury 1964
123. *Peperomia leptostachya*	—	Juice from leaves used for burns and skin and eye infections	Lewis and Elvin-Lewis 1977
124. *Peperomia obtusifolia* Dietr. cv. Aurea	Alkaloids (?)	Toxic (see table 5.1)	Der Marderosian, Giller, and Roia 1976
125. *Philodendron cordatum*	—	Mixed with soap for treating eczema and other skin conditions	Lewis and Elvin-Lewis 1977
126. *Philodendron oxycardium* Schott	—	Toxic (see table 5.1)	Der Marderosian, Giller, and Roia 1976
127. *Philodendron sagittifolium* Liebm.	—	Nontoxic (see table 5.1)	*Ibid.*
128. *Philodendron* spp.	Calcium oxalate, irritant principles	Poisoning in cats with debilitation and listlessness and complete destruction of kidney function with no pain; leaves and stems dangerous if eaten in quantity	Kingsbury 1964; Hardin and Arena 1974; Lewis and Elvin-Lewis 1977
129. Same	Calcium oxalate	Two infant ingestions; no data	O'Leary 1964
130. *Pilea cadierei* Gagnep. & Guillaumin	Alkaloids (?)	Toxic (see table 5.1)	Der Marderosian, Giller, and Roia 1976
131. *Podophyllum peltatum* L.	—	Toxic (see table 5.1)	*Ibid.*

No. / Species	Active compounds	Toxicity / Effects	Reference
132. Same	16 physiologically active compounds found in 2 groups, lignans and flavonols	Resin is violent cathartic; resin used in treatment (topically) of venereal warts; resin cytotoxic; handling rhizome caused severe conjunctivitis, keratitis, and ulcerative skin lesions; ripe fruit least toxic, but occasional catharsis with unripe fruit	Kingsbury 1964; Hardin and Arena 1974; Lewis and Elvin-Lewis 1977
133. Same	—	One ingestion; no data	Watt and Breyer-Brandwijk 1962
134. Same	—	Poisoning with resin, with complete recovery	Balucani and Zellers 1964
135. Same	—	Fatal poisoning with resin	Kaymakcalan 1964
136. *Saintpaulia ionantha* Wendl. cv. Ocean Waves	Alkaloids (?)	Toxic (see table 5.1)	Der Marderosian, Giller, and Roia 1976
137. *Saintpaulia* spp.	—	Two infant ingestions of leaves; no data	O'Leary 1964
138. *Sansevieria thyrsiflora* Thunb.	—	Toxic (see table 5.1)	Der Marderosian, Giller, and Roia 1976
139. Same	—	Leaf juice used for earache and toothache; root chewed and juice swallowed for hemorrhoids and intestinal worms; juice kills intestinal parasites	Lewin 1962; Watt and Breyer-Brandwijk 1962
140. *Sansevieria trifaciata* Praim.	Hemolytic saponin and organic acids	No toxicity data given	Watt and Breyer-Brandwijk 1962
141. *Sansevieria trifaciata* Praim. var. *laurentii*	—	Toxic (see table 5.1)	Der Marderosian, Giller, and Roia 1976
142. *Selaginella pallescens* Presl.	—	Toxic (see table 5.1)	Ibid.

Table 5.2 continued

Plant	Principles Reported	Toxicity/Edibility Data	Literature Reference
143. *Selaginella rupestris* Spreng.	—	Smoked with *Lycopodium clavatum* for headache; powdered rhizome rubbed into venereal sores	Watt and Breyer-Brandwijk 1962
144. *Solanum pseudocapsicum* L.	Alkaloids	Toxic (see table 5.1)	Der Marderosian, Giller, and Roia 1976
145. Same	Solanaceous alkaloids solanine, solanidine, solanocapsine	Established reputation for toxicity but no clear cases recorded; fruit (3 to 4 berries) said to cause fatal poisoning in children; unripe fruit most toxic	Kingsbury 1964; Hardin and Arena 1974; Watt and Breyer-Brandwijk 1962
146. Same	—	Eight ingestions of fruits recorded in children; no data	O'Leary 1964
147. *Syngonium podophyllum* Schott var. *albo-virens*	—	Toxic (see table 5.1)	Der Marderosian, Giller, and Roia 1976
148. *Xanthosoma hoffmannii* Schott	—	Nontoxic (see table 5.1)	*Ibid.*
149. *Xanthosoma* spp.	Irritant juice	Dermatitis possible; tubers edible after cooking	Kingsbury 1964; Lewis and Elvin-Lewis 1977; Morton 1958

fenbachia picta and *D. exotica* in the albino Wistar rat. They found that the juice elicited an oral swelling reaction very similar to that observed in humans. A single-dose oral toxicity determination in rats showed an LD_{50} over 160 ml/kg of the juice (whole plant). An eye irritation study in albino rabbits revealed principally a conjunctival involvement in addition to some reversible damage to the cornea. These researchers ran an oxalic acid determination on the whole plant and found that the juice of *D. exotica* gave values of 0.37 percent and 0.15 percent, respectively. *Dieffenbachia picta* gave values of 0.21 percent and 0.10 percent. These represented total oxalate contained in the plants, whether it was there as calcium oxalate crystals or as other oxalate salts. They also determined blood histamine concentrations in rats treated with the juice of *D. picta*, and the levels were significantly increased over those of untreated animals. Histological study of tongue tissue from animals treated with *Dieffenbachia* juice showed edema, vascular congestion, degeneration of the basement membrane, and an inflammatory reaction. Fochtman and colleagues found that trypsin digestion of the juice decreased the toxicity observed in the rat mouth. This lends credence to the concept that an enzyme is involved. Further, they note that rats pretreated with an antihistamine (diphenhydramine) histologically showed some protection against the effects of the juice, but treatment with cortisone acetate only delayed the reaction. These authors attribute the toxicity of the juice to a labile "proteinlike" substance, not to the oxalate content, as was previously believed. They claim that the mechanism of toxicity seems to be associated with histamine release. Curiously, Walter and Khanna (1972) did not refer to the earlier studies of Fochtman and associates (1969).

The most recent study found on *Dieffenbachia* was that by Ladeira, Andrade, and Sawaya (1975). This group studied the toxic effects of *D. picta* on guinea pigs. Administration of juice of this plant orally to the guinea pig showed that the toxic component (LD_{50} between 600 and 900 mg of stem/animal in 24 hr) is contained in the stem and is unstable on heating to 100°C or on vacuum drying at 50–60°C. Curiously, administration of the antihistamine chlorpheniramine or incubation of the juice with chymotrypsin did not hinder edema formation in the mouths of test animals. Perhaps the use of different antihistamine or a different digesting enzyme made a difference. Further, these investigators found that stem juice is also toxic when injected i.p. (LD_{50}, 1 g) but not when administered directly into the stomach by intubation. Stomach acid or intestinal enzymes may have something to do with explaining this. Also, they found that the precipitate of the stem juice obtained by centrifugation at 3°C, 10,000 rpm for 1 hr, but not the supernatant, contains the toxic component. The supernatant contains one substance that causes contraction of smooth musculature. They concluded that it is probably not histamine, since dosing with an antihistamine compound (promethazine) did not inhibit its effect. Leaves also contain this component, but they did not exhibit any toxic effect when tested orally in animals. No nitrogen was detected in the residue of the stem juice. Thus, these studies confirm that the leaf of D. picta is very much less active than the stem. Greater toxicity by i.p. administration versus gastric intubation is found. The supernatant of the stem juice did not provoke toxic effects. The lack of nitrogen in the toxic residue of the stem juice and its lack of digestability speak against the toxic protein hypothesis. Suffice it to say that much more work is needed to resolve the question of the nature of the toxic principle(s) in Dieffenbachia.

Digitalis purpurea (foxglove) is well acknowledged to be a toxic garden plant. However, very few incidences of human toxicity from the plant are recorded. As indicated in table 5.2, most problems here relate to overdose of the drug Digitalis or its isolated glycosides (digitoxin, etc.), which are widely used as cardiotonics.

Dracaena sanderiana has no direct reference in the literature relating to toxicity or edibility. Screening has shown toxicity in rats on i.p. administration (Der Marderosian, Giller, and Roia 1976). Several species were found that had various medical and folkloric uses. These include the use of Dracaena resin for stopping hemorrhages, the juice for relief of postparturient pain, a root-derived remedy for rheumatism, and a leaf infusion for skin diseases, all noted in table 5.2. Two species, Dracaena cylindrica and Dracaena laxissima, have poisonous qualities.

Over the years there has been considerable controversy over Euphorbia pulcherrima (poinsettia) as to whether or not it is really toxic. The various references cited in table 5.2 show no toxicity of the flowers, bracts, and leaves in mice (oral administration) (Der Marderosian, Giller, and Roia 1976); the death of a 2-year-old child after eating a poinsettia leaf (Kingsbury 1964), from a single unverifiable case reported in Hawaii in 1919; and several instances of human toxicities usually manifested as local irritation (abdominal pain, with vomiting and diarrhea). Chemical studies have revealed that poinsettia contains several common plant sterols and triterpenes, none of which is commonly considered toxic (Dominguez et al. 1967). However, this study was oriented toward looking for compounds with possible hormonal activity in different parts of the plant. More chemical studies need to be carried out on the locally irritating substances following activity by some bioassay procedure. Stone and Collins (1971) fed large amounts of poinsettia leaf, bract, and flower homogenate to 160 rats (141 females, 19 males) orally and found no signs of toxicity or ill effects in these animals. Doses as high as 50 g per kg gave zero mortality. We have carried out more recent studies in mice, rats, rabbits, and a dog that show that, orally, large amounts of poinsettia are not toxic in these animals (Der Marderosian, unpublished data). Perhaps, as discussed by Kingsbury (1975), horticultural manipulation to breed showier, longer-lasting and differently colored poinsettias has affected the nature of the toxic principles in some way. Once again attention must be paid to careful botanical identification of species and the various varietal forms used in any toxicological study. Recently D'Arcy (1974) reported on a severe contact dermatitis resulting from poinsettia. It occurred in a 66-year-old man who cut and bundled large (1 to 2 meters tall) varieties of poinsettia with red and white bracts and carried them against his bare chest on a hot day (82°F). After a day or so, he sustained itching and soreness under his watchband. Three days after contact, he was covered over hands, arms, chest, and shoulders with erythema and some localized scaling. After two more days, erythema and itching occurred on the thighs, followed by papules and vesicles. The condition persisted for a long time. The delay before the appearance of symptoms is suggestive of allergenic hypersensitivity. D'Arcy notes that little has been written indicating the potential of poinsettia as a contact allergen or irritant.

In the case of *Euphorbia tirucalli* (pencil bush), a fair number of references were found that document the irritancy of its profuse latex (see table 5.2). The stem sap showed no toxicity i.p. in rats (Der Marderosian, Giller, and Roia 1976), but several articles indicate its ability to cause dermatitis and severe irritation to throat, mouth, and stomach if swallowed. One case is given of an adult male in Africa who died from hemorrhagic gastroenteritis after swallowing the latex as a cure for sterility (Watt and Breyer-Brandwijk 1962). A fairly detailed study of the effect of the latex of candelabra cactus (*Euphorbia lactea*) and pencil bush or tree (*E. tirucalli*) has been published by Crowder and Sexton (1964). Both were shown experimentally to produce keratoconjunctivitis and uveitis in dog eyes. Histological sections of these revealed that the corneal opacity produced by the latex of these plants tended to clear without sequelae. Vascularization of the peripheral stroma and polymorphonuclear leukocyte infiltration were a consistent histological finding. The human case that prompted this study occurred in a 42-year-old white male who was cutting candelabra cactus in the Florida Keys. This and similar cases reveal that the typical symptoms include immediate burning of the eyeball and eyelids, tearing, and photophobia. In 8–12 hours chemosis of lids and conjunctiva usually follow, accompanied by blurred vision and increased pain. Treatment included use of a local anesthetic, an antiinflammatory steroid (dexamethasone), and an antibiotic (neomycin).

Reference to *Ficus elastica* (*decora*) (rubber plant) and *Ficus lyrata* (*pandurata*) in table 5.2 reveals that, in the former, leaf and stem caused no toxicity in rats and mice. The latter showed leaf toxicity in rats (i.p.) and no toxicity with the unripe fruits. Reference to studies of related species shows that many have edible fruits and a latex with proteolytic properties (ficin). This latter property has been put to use in treating intestinal worms. The rubber plant has been reported to be ingested by young children in this country, but no toxic properties have been reported, while several species of *Ficus* are known to be toxic (O'Leary 1964).

In the case of *Gladiolus* species, the toxicological literature is once again inconclusive. The flowers of *G. gandavensia* are toxic to rats on i.p. administration (Der Marderosian, Giller, and Roia 1976). Other species have medicinal uses, and in some cases the cooked corm is edible.

Hemerocallis lilio-asphodelus (*flava*) L. (day lily) has flowers, leaves, and roots that are nontoxic to mice (Der Marderosian, Giller, and Moia 1976). The flowers are edible (O'Leary 1964).

Hippeastrum species generally contain several active alkaloids, with the bulbs being reported as toxic to humans (Lewis and Elvin-Lewis 1977; Morton 1962). Typical symptoms include gastrointestinal upset leading to convulsions and death. The leaf of *Hippeastrum* hybrid (amaryllis) was nontoxic to mice (p.o.), but the flowers were toxic to rats (i.p.) (Der Marderosian, Giller, and Roia 1976).

The *Hydrangea* species generally showed a toxicity profile, with several instances of cyanogenic glycosides in leaves and buds being reported. Variable results are seen in the case of mice and rat toxicity data.

Lantana camara L. cv. Gold mound showed uniform toxicity of its ripe and unripe fruits to rats on i.p. administration (Der Marderosian, Giller, and Roia 1964). In human experiences the unripe fruits have poisoned children and have caused at least one death (Wolfson and Solomons 1964).

The symptoms resemble those of belladonna poisoning. The active principles are pentacyclic triterpenes (lantadene A and B). These are hepatogenic photosensitizers, also causing gastroenteritis in grazing animals.

Little data could be found on the prayer plant (*Maranta leuconeura*), but the variety studied showed leaf toxicity to rats on i.p. administration (Der Marderosian, Giller, and Roia 1976).

The "cut-leaf" philodendron (*Monstera deliciosa*) showed toxicity (leaf) to both rats and mice (Der Marderosian, Giller, and Roia 1976). Other studies reveal the presence of calcium oxalate and irritant juice (Kingsbury 1964). The fully ripe fruit (ceriman) is a pleasant, edible conelike fruit (Morton 1962).

Nothing could be found on the toxicity of the Boston fern (*Nephrolepis exaltata*), as is shown in table 5.2. The variety studied was not toxic to rats or mice (Der Marderosian, Giller, and Roia 1976).

The positive-toxic control throughout this study was *Nerium oleander*. Reference to table 5.2 reveals a large body of data on this plant. The red, double, pink, and white varieties were all shown to be uniformly toxic to both mice and rats (Der Marderosian, Giller, and Roia 1976). The toxicity is directly related to the cardioactive glycosides oleandroside and nerioside. Their symptomatology is given in table 5.2.

Paeonia lactiflora stem, leaf, and flowers were found to be nontoxic to mice (Der Marderosian, Giller, and Roia 1976). No other data could be located on the possible toxicity of the genus. Likewise, information was lacking on *Peperomia obtusifolia* (see table 5.2). Toxic effects were observed in both rats and mice when fed leaf material (Der Marderosian, Giller, and Roia 1976). The juice and leaves of *Peperomia* species have been used for burns and skin and eye infections (Sturtevant 1919).

A paucity of reliable toxicological information on *Philodendron* was also noted. *P. oxycardium* leaf and stem showed toxicity in rats and mice (Der Marderosian, Giller, and Roia 1976). On the other hand, *P. sagittifolium* showed no toxicity in mice (leaf and immature flowers), as may be seen in table 5.2. In general, philodendrons are considered to contain calcium oxalate and irritating principles. Each year there are many cases of cats being poisoned by these plants, but no definitive studies have been carried out on the genus with respect to active principles and their full toxicological properties.

No information could be found for *Pilea cadierei* (aluminum or water-

melon plant). The leaf and stem were noted to be toxic to rats on i.p. administration (Der Marderosian, Giller, and Roia 1976).

Podophylum peltatum (mayapple or American mandrake) represents an
example of one plant that has been well studied. At least sixteen physiologically active compounds have been noted in the crude resin podophyllin (see
table 5.2). The resin is a well-known cathartic and has cytotoxic properties.
It is used in the treatment of venereal warts and certain skin cancers. Ca-
tharsis results from ingestion of the unripe fruit. The ripe fruit is least
toxic and is considered by some to be edible. The ripe and unripe fruits
are toxic to rats on i.p. administration (Der Marderosian, Giller, and Roia
1976).

The African violets (*Saintpaulia* spp.) seem to be generally nontoxic. No
toxicity is seen in rats (see table 5.2), and reports on ingestions by infants
reveal no toxic effects. The snake plant (*Sansevieria trifaciata*), both leaf and
flowers, was shown to be toxic to rats (i.p.) and mice (p.o.). This plant is
considered to contain hemolytic saponins and organic acids. One species
has reported medicinal uses. Nothing could be found on *Selaginella pallescens*, but the leaf proved toxic to rats on i.p. administration (Der Marderosian, Giller, and Roia 1976).

The genus *Solanum* is well known for its toxic effects; however, several
species need further clarification as to how toxic they are and what parts are
nontoxic. In this study, extracts of *Solanum pseudocapsicum* L. (Jerusalem
cherry) showed the presence of alkaloids, and while the fruits (ripe and
unripe) were toxic to rats (i.p.), the leaf and ripe fruits were nontoxic to
mice (p.o.) (Der Marderosian, Giller, and Roia 1976). Otherwise, this plant
has a long-established reputation for toxicity, but few clear cases of poisoning are on record (Kingsbury 1964; Lewis and Elvin-Lewis 1977; Watt and
Breyer-Brandwijk 1962) (see table 5.2). The unripe fruit is considered most
toxic. Several ingestions of fruits are recorded in children (O'Leary 1964).

Practically nothing of a toxicological nature could be found in the literature on *Syngonium* and *Xanthosoma*. Table 5.2 shows that the former has
some toxicity while the latter has none. Some species of *Xanthosoma* have
irritant juices, and dermatitis is thus possible. The tubers are considered
wholesome after cooking and are an important vegetable in tropical countries.

In summarizing this attempt to combine preliminary acute toxicity studies
and literature information on household plants, it may be said that much

work remains to be done. At this point several species in table 5.2 may be considered relatively innocuous, therefore precluding the necessity of gastric lavage if they are accidentally ingested. Others remain in the potentially toxic category, and many with known toxicities have been confirmed. The development of toxicity profiles is a long and tedious task. Short of direct experimentation with humans, which is difficult to justify, we must continue to rely on the literature as well as testing in several animal species. This must be followed by relating the data to any incidence of accidental ingestion in order to develop proper clinical management of cases in humans. This multiphasic approach will ultimately provide quantitative data that can be applied to any case encountered. However, we will be plagued for a long time with the following problems (Kingsbury 1964, 1975; Der Marderosian 1966):

1. Lack of proper identification of plant as to correct species, variety, etc.
2. Continued use of popular names for plants that are not specific enough for toxicological purposes.
3. Continued horticultural manipulation of household ornamentals, with possible subsequent development of plant varieties having different pharmacological properties.
4. Lack of knowledge as to quantity of plant material ingested.
5. Lack of information related to degree of maturity of plant or plant parts ingested. Many plants are quite complex biochemically and often contain numerous active substances in differing concentrations at various stages of growth.
6. Lack of information as to whether a hard seed coat has been pierced or cracked or softened prior to ingestion. Many seeds can pass through the alimentary tract intact because of a hard seed coat.
7. Inability to determine whether toxic plant principles have been absorbed. Gastrointestinal tract lesions may allow usually nontoxic materials (e.g., saponins) to be absorbed.
8. Lack of knowledge related to general health, age, and sensitivity of the person to plant poisons.
9. Lack of information on any special inherited susceptibility and blood composition of the person ingesting a plant material (e.g., favism).
10. Lack of information relating to how the toxic principles of a plant

may be altered (boiled, pickled, dried, cooked, etc.) prior to ingestion. Many toxic plants can be rendered nontoxic by certain treatments.

11. Lack of a system for marshaling the necessary specialists to fully han-
dle and properly record all plant poisonings.

12. Lack of a central national plant information center that can provide
data on the toxicity and/or edibility of various species.

Clinical Management of Plant Poisoning

At this point it would be instructive to review the various clinical manage-
ments developed for handling poisoning by common household plants. Obviously, it will be impossible to give all the necessary details, but there are
sources available. These include the National Clearinghouse for Poison
Control Centers, National Institutes of Health, Washington, D.C., which
maintains index cards and publishes a bulletin on references, statistics, and
occasionally specific information on plant poisonings and their management. Poisindex is another source of specific poison management information. It is a bank of computer-generated information on compounds (including plants) and their corresponding treatments in case of poisoning. One
subscribes to the service and is provided with a notebook-sized film file of
microfiche cards, each containing over 2,000 entries. These are alphabeti-
cally arranged and can be read in a special viewer. Colored pictures of plants
and mushrooms are also provided. Poisindex is trademarked under Micro-
medix, Inc., which is associated with the National Center for Poison Information, 2645 So. Santa Fe Drive, Denver, Colorado 80223.

With respect to plants in general, it should be emphasized that prompt removal of anything ingested is of prime importance. Once material is absorbed, treatment must proceed along symptomatic lines. Ipecac syrup (not
the fluid extract or elixir) should be given to induce emesis, with a 30 ml
dose for adults and a 15 ml dose for children. This may be repeated once.
Following this, fluids should be forced orally, the patient kept walking, and
manual stimulation of the pharnyx occasionally tried to hasten emesis. Caution should be used not to give too much ipecac syrup, since the emetine in
it can cause cardiac toxicity. For material that has been absorbed, activated
charcoal (Norit-AR) may be used at a level of 5–10 times the approximate
dose of the ingested plant material. For an average adult 20–30 grams is

sufficient. A water slurry of the charcoal flavored with something sweet will make it palatable. One so-called universal antidote containing tannic acid is no longer recommended, since its active principle, tannic acid, may itself be toxic (Oler, Neal, and Mitchell 1976). Also, activated charcoal can inactivate ipecac and thus should not be used concomitantly. Charcoal will not work in the case of cyanide poisoning. Cyanide poisoning by cyanogenic glycosides (e.g., amygdalin in members of the Rosaceae) should be managed using a commercial cyanide antidote kit. These kits provide specific details and usually include amyl nitrite, sodium nitrite, and sodium thiosulfate. The sodium nitrite converts some of the blood hemoglobin to methemoglobin, which in turn combines preferentially with cyanide in competition with the respiratory enzyme cytochrome oxidase. Sodium thiosulfate converts some cyanide (including that from dissociation of cyanmethemoglobin) into the more stable and less toxic thiocyanate.

In the case of poisoning by *Solanum pseudocapsicum* (Jerusalem cherry), which contains the solanine-type alkaloids (e.g., solanocapsine), one should watch for the characteristic gastrointestinal irritation. Other typical symptoms include headache, nausea, depression, bradycardia, dyspnea, sweating, muscle weakness, salivation, and renal failure. Atropine-like effects are also possible, since atropine-type compounds are also present in the plant. These may be seen in the urine through clinical laboratory monitoring. Management includes such measures as maintenance of respiration (artificially if needed), control of convulsions (i.v. diazepam), induction of emesis through lavage or with ipecac, use of activated charcoal, administration of saline laxatives (e.g., magnesium or sodium sulfate) to hasten gastrointestinal motility and evacuation, use of i.v. fluids for control of electrolyte problems precipitated by severe gastrointestinal symptoms; and use of physostigmine if the atropine anticholinergic-type reaction predominates (Alexander, Forbes, and Hawkins 1948).

The ornamental plants that contain cardioactive glycosides of the digitalis (*Digitalis purpurea*) type may generally be treated the same way that digitalis overdose is managed. These include *Convallaria majalis* (lily-of-the-valley) and *Nerium oleander* (oleander). Clinically, intoxication with these causes a marked bradycardia with varying degrees of heart block. Vomiting is seen shortly after ingestion and may continue for several hours, getting worse as the heart muscle deteriorates over the 6 to 24 hours following ingestion. For laboratory assessment careful measurement of serum potassium is important, since hyperkalemia usually follows ingestion of cardioactive glycosides.

Treatments include establishment of respiration (artificial if needed), prevention of absorption (emesis induction, gastric lavage, activated charcoal, saline laxatives), maintenance of proper potassium levels, use of a transvenous electrical pacer if the patient is severely poisoned, use of low-dose diphenylhydantoin to improve atrial ventricular conduction and help terminate complete heart block as well as help heart rate, and use of atropine to restore atrial activity and raise heart rate. Agents such as sodium EDTA, magnesium salts, and propranolol are of dubious value in cases of massive overdose. Potassium salts are contraindicated unless there is a demonstrated hypokalemia in acute massive cardiac glycoside overdose (Rumack, Wolfe, and Gilfinche 1974). Szabuniewicz, McCrady, and Camp (1971) have found atropine used in conjunction with propranolol to be beneficial in the treatment of experimentally induced oleander poisoning in the dog.

The group of plants in the Araceae (viz., *Dieffenbachia, Arisaema, Caladium, Philodendron,* and *Xanthosoma*), also known as aroids, can perhaps be treated in a similar fashion as far as poisoning is concerned. General treatment includes maintenance of respiration; treatment of convulsions with i.v. diazepam; removal of plant material via emesis or lavage (except if swelling is severe); use of milk, water, or antacids to dilute the calcium oxalate and to flush out and soothe the oral pharynx; use of analgesics (e.g., meperidine) for pain; possible use of antihistamine or corticosteroids, although their effectiveness is equivocal; and maintenance of adequate hydration via intravenous fluids (Fochtman et al. 1969; Pohl 1961). Although quite experimental, perhaps papain should be tried as a proteolytic enzyme to attempt to degrade the "dumbcain" enzyme, if it in fact is the major active principle. Papain is quite nontoxic orally.

The only clinical management of poisoning by the *Euphorbia* species published relates to transfer of the latex to the eyes. As reported by Crowder and Sexton (1964), these species are capable of causing erosion of the corneal epithelium, marked chemosis and injection of the conjunctiva, erythema and chemosis of the lids, decreased visual acuity, and corneal edema. Treatments suggested include use of a local anesthetic (tetracaine) for the pain; a corticosteroid, dexamethasone, for the inflammation; and an antibiotic, neomycin, for any possible infection. Oral poisoning would probably be best managed by the procedures suggested for the aroids, namely, removal of plant material via emesis or lavage; use of milk, water, or antacids to dilute, flush out, and soothe irritated oral membranes; and symptomatic treatment.

Clinical management is probably not necessary for accidental ingestion of small amounts of *Anthurium, Begonia, Ficus, Nephrolepis, Paeonia, Peperomia,* and *Saintpaulia,* as shown in table 5.2. The remainder require more accidental ingestion experience or toxicological assessment in several animal species before confident clinical management can be suggested. As with all medical information, changes are made as more data becomes available. There, no doubt, will be alterations in the clinical management procedures suggested here as time goes on. However, it is anticipated that the ideas presented here will be useful for the present.

REFERENCES

Alexander, R., G. Forbes, and E. Hawkins. 1948. "A Fatal Case of Solanine Poisoning." 1948. *Br. Med. J.* 2:518–19.

Allen, P. 1943. *Poisonous and Injurious Plants of Panama.* Baltimore: Williams and Wilkins.

Arnold, H. 1944. *Poisonous Plants of Hawaii.* Honolulu: Tongg.

Balucani, M. and D. Zellers. 1964. "Podophyllum Resin Poisoning with Complete Recovery." *J. Am. Med. Assoc.* 189:639–40.

Burkill, I. 1935. *Dictionary of the Economic Products of the Malay Peninsula.* London: Crown Agents.

Chopra, R., R. Badhwar, and S. Ghosh. 1965. *Poisonous Plants of India.* New Delhi: Indian Council of Agricultural Research Service.

Council of Scientific and Industrial Research. 1948. "Begonia Linn. Begoniaceae." In *The Wealth of India: Raw Materials,* vol. 1. New Delhi.

Crowder, J. and R. Sexton. 1964. "Keratoconjunctivitis Resulting from the Sap of Candelabra Cactus and the Pencil Tree." *Arch. Ophthalmol.* 72: 476–84.

Dahlgren, B. 1947. *Tropical and Subtropical Fruits.* Chicago: Chicago Natural History Museum.

Dalziel, J. 1948. *Useful Plants of West Tropical Africa.* London: Crown Agents.

D'Arcy, W. 1974. "Severe Contact Dermatitis from Poinsettia." *Arch. Dermatol.* 109:909–10.

Der Marderosian, A. 1966. "Poisonous Plants In and Around the Home." *Am. J. Pharm. Educ.* 30:115–40.

Der Marderosian, A., F. Giller, and F. Roia, Jr. 1976. "Phytochemical and Toxicological Screening of Household Plants Potentially Toxic to Humans. 1." *J. Toxicol. Environ. Health.* 1:939–53.

Der Marderosian, A. Unpublished results.

Dominquez, X, J. Delgado, M. Maffey, J. Mares, and C. Rombold. 1967. "Chemical Study of the Latex, Stem, Bracts, and Flowers of 'Christmas Flower' (*Euphorbia pulcherrima*). 1." *J. Pharm. Sci.* 56:1184–85.

Fochtman, F., J. Manno, C. Winek, and J. Cooper. 1969. "Toxicity of the Genus *Dieffenbachia.*" *Toxicol. Appl. Pharmacol.* 15:38–45.

Hardin, J. and J. Arena. 1974. *Human Poisoning from Native and Cultivated Plants,* 2d ed. Durham, N.C.: Duke University Press.

HEW. 1976. "Tabulations of 1974 Case Reports." In *National Clearinghouse for Poison Control Centers Bulletin,* September 1976. Bethesda, Md.: United States Department of Health, Education and Welfare.

Irvine, F. 1961. *Woody Plants of Ghana.* London: Oxford University Press.

Kalaw, M. and F. Sacay. 1925. "Some Alleged Philippine Poisonous Plants." *Philipp. Agric.* 14:421–27.

Kaymakcalan, S. 1964. "Fatal Poisoning with Podophyllum Resin." *J. Am. Med. Assoc.* 190:558.

Kingsbury, J. 1964. *Poisonous Plants of the United States and Canada.* Englewood Cliffs, N.J: Prentice-Hall.

—— 1969. "Phytotoxicology I. Major Problems Associated with Poisonous Plants." *Clin. Pharmacol. Ther.* 10:163–69.

—— 1975. "Phytotoxicology." In L. Casarett and J. Doull, eds., *Toxicology, the Basic Science of Poisons,* pp. 591–603. New York: Macmillan.

Ladiera, A., S. Andrade, and P. Sawaya. 1975. "Studies on *Dieffenbachia picta* Schott: Toxic Effects in Guinea Pigs." *Toxicol. Appl. Pharmacol.* 34:363–73.

Lewin, L. 1962. *Gifte und Vergiftungen, Lehrbuchs der Toxicologie,* 5th ed. Ulm/Donau, W. Germany: K. F. Haug Verlag.

Lewis, W. and M. Elvin-Lewis. 1977. *Medical Botany.* New York: Wiley.

Morton, J. 1958. "Ornamental Plants with Poisonous Properties." *Proc. Fl. State Hortic. Soc.* 71:372–80.

—— 1962. "Ornamental Plants with Toxic and/or Irritant Properties. 2." *Proc. Fl. State Hortic. Soc.* 75:484–91.

O'Leary, S. 1964. "Poisoning in Man from Eating Poisonous Plants." *Arch. Environ. Health.* 9:216–42.

Oler, A., M. Neal, and E. Mitchell. 1976. "Tannic Acid: Acute Hepatotoxicity Following Administration by Feeding Tube." *Food Cosmet. Toxicol.* 14:565–69.

Opp, M. 1977. "Beautiful but Dangerous." *Med. World News,* May 16:38–43.

Pohl, R. 1961. "Poisoning by *Dieffenbachia.*" *J. Am. Med. Assoc.* 177:812.

Quisumbing, E. 1951. *Medicinal Plants of the Philippines.* Manila: Philippine Department of Agriculture and Natural Resources.

Rumack, B., R. Wolfe, and H. Gilfinche. 1974. "Diphenylhydantoin (Phentoin) Treatment of Massive Digoxin Overdose." *Br. Heart J.* 36:405–8.

Schenk, G. 1955. *The Book of Poisons.* New York: Rinehart.

Spector, W., ed. 1955. *Handbook of Toxicology,* Tech. Rep. 55-16, vol. 1, pp. 74–77. Wright-Patterson AFB, Ohio: Wright Air Development Center.

Stone, R. and W. Collins. 1971. "*Euphorbia pulcherrima:* Toxicity to Rats." *Toxicon* 9:301–2.

Sturtevant, E. 1919. *Sturtevant's Notes on Edible Plants, New York.* Geneva, N.Y.: Agricultural Research Station.

Szabuniewicz, M., J. McCrady, and B. Camp. 1971. "Treatment of Experimentally Induced Oleander Poisoning." *Arch. Pharmacodyn. Ther.* 189:12–21.

Terra, G. 1966. *Tropical Vegetables.* Amsterdam: Royal Tropical Institute and Netherlands Organization for International Assistance.

Walter, W. and P. Khanna. 1972. "Chemistry of the Aroids I. *Dieffenbachia seguine, amoena* and *picta.*" *Econ. Bot.* 26:364–72.

Watt, J. and M. Breyer-Brandwijk. 1962. *The Medicinal and Poisonous Plants of Southern and Eastern Africa.* Edinburgh and London: E. and S. Livingstone.

Webb, L. 1948. *Guide to Medicinal and Poisonous Plants of Queensland,* p. 20. Melbourne: Council for Scientific and Industrial Research.

Wolfson, S. 1964. "Acute Poisoning in Children from Ingestion of the Green Fruit of *Lantana camara.*" In *National Clearinghouse for Poison Contgrol Centers Bulletin,* January-February 1964. Washington, D.C.: National Institutes of Health.

Wolfson, S. and T. Solomons. 1964. "Poisoning by Fruit of *Lantana camara.*" *Am. J. Dis. Child.* 107:173–76.

6 Cocarcinogenic Irritant Euphorbiaceae

A. Douglas Kinghorn

The Euphorbiaceae, or spurge family, consists of over 300 genera and 7,000 species (Webster 1967). Although plants of this family have important economic uses as foodstuffs, as medicinals, and in industry, particularly as sources of rubber and timber, a great deal of scientific interest has focused on the piscicidal, purgative, and irritant principles of several species. The attempts of over 150 years to establish the cathartic principles of *Croton tiglium* were given a new impetus when Berenblum discovered the tumor-promoting (cocarcinogenic) activity of croton oil (Berenblum 1941). These efforts culminated with the isolation of the irritant cocarcinogenic factors of the oil as esters of the tetracyclic diterpenoid phorbol (Hecker 1968). Phorbol esters, which are now a valuable biochemical tool in the study of skin carcinogenesis at the cellular level, promote sarcomas in mouse skin previously treated with a subcarcinogenic dose of a carcinogen (Boutwell 1974). In the decade since the structure elucidation of phorbol (Hoppe et al. 1967; Pettersen et al. 1967), many related diterpene esters have been isolated from the spurge family, especially from the genus *Euphorbia* (Hecker 1977).

The first part of this chapter will review the chemical types of biologically active diterpenoids that have been recently isolated from the Euphorbiaceae. Botanical characteristics and examples of the economic uses of irritant members of the spurge family will be considered, with an attempt to relate the distribution of species containing irritant diterpenoids to existing taxonomic groups. Studies on diterpenoids in two-stage carcinogenesis experiments and structural factors affecting their toxicity will be described. Finally, the health hazard that these plants present to humans and livestock will be discussed.

Cocarcinogenic Irritant Diterpenoids of the Euphorbiaceae

Phorbol Esters

The isolation of pure biologically active constituents from *Croton tiglium* was achieved by two independent groups (Hecker, Bresch, and Meyer 1965; Van Duuren and Orris 1965). A detailed description of the painstaking fractionation procedures applied by the Hecker group at Heidelberg has been published (Hecker and Schmidt 1974). Phorbol, and the major croton oil diester, 12-O-tetradecanoylphorbol-13-acetate (TPA), are shown in figure 6.1.

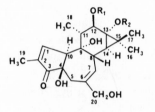

PHORBOL $R_1 = R_2 = H$

TPA (12-O-tetradecanoyl-phorbol-13-acetate)

$R_1 = CO(CH_2)_{12}CH_3$; $R_2 = COCH_3$

Figure 6.1. The structure of phorbol and TPA.

Esters of the diterpene alcohol, phorbol, have been found in only a few species in the plant kingdom (see figure 6.2). In croton oil, 50 percent of the phorbol esters are extracted into hydrophilic solvents (phorbol 12,13-diesters), and the remainder are extracted into hydrophobic solvents (phorbol 12,13,20-triesters) (Hecker and Schmidt 1974). Altogether, 14 phorbol diesters have been isolated from *Croton tiglium*; they have been divided into two groups, factors A and B, by Hecker and coworkers. A factors are long-chain 12-O-acylates, with a short acyl chain at C-13, while B factors have the opposite ester functionality (Hecker and Schmidt 1974) (see figure 6.2). Several phorbol 12,13-diesters have been isolated from the ornamental pencil tree, *Euphorbia tirucalli* (Fürstenberger and Hecker 1977a), and a pis-

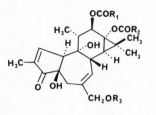

Plant	R_1	R_2	R_3	
Croton tiglium	Long alkyl	Short alkyl	H	(A Factors)
	Short alkyl	Long alkyl	H	(B Factors)
Euphorbia tirucalli	Complex alkyl	CH_3	H	
Sapium japonicum	Deca-2,4,6-trienoate	CH_3	H	
Croton sparsiflorus	$(CH_2)_{10}CH_3$	CH_3	Linolenate	
Euphorbia franckiana, E. coerulescens	$CH(CH_3)_2$	CH_3	Angelate	

Figure 6.2. Some phorbol esters isolated from the Euphorbiaceae.

cicidal 12,13-diester of phorbol was obtained from *Sapium japonicum* (Ohigashi et al. 1972) (see figure 6.2).

Phorbol triesters have been termed "cryptic" cocarcinogens, since upon removal of the 20-O-acyl group by selective hydrolysis cocarcinogenic phorbol 12,13-diesters are produced (Hecker 1976). Examples of phorbol triesters have been reported in *Croton sparsiflorus* (Upadhyay and Hecker 1976) and in *Euphorbia franckiana* and *E. coerulescens* (Evans 1977) (see figure 6.2).

Figure 6.3 shows the substituted phorbol alcohols, also based on the hydrocarbon tigliane, that are produced upon hydrolysis of the biologically active esters; it includes all that have been reported to date. Within a group of almost 60 *Euphorbia* species, Evans and Kinghorn (1977) showed that 12-deoxyphorbol is the most widely distributed tigliane alcohol. Irritant and/or cocarcinogenic esters of 12-deoxyphorbol have been reported in *Euphorbia triangularis* (Gschwendt and Hecker 1974), *E. fortissima* (Kinghorn and Evans 1975a), *E. coerulescens* and *E. polyacantha* (Evans and Kinghorn 1975a), *E. helioscopia* (Evans, Kinghorn, and Schmidt 1975), *E. resinifera* (Hergenhahn, Kusumoto, and Hecker 1974), *E. poisonii* (Evans and Schmidt 1976; Schmidt and Evans 1976, 1977a), and *E. unispina* (Schmidt and Evans 1976, 1977a). Esters of 12-deoxy-16-hydroxyphorbol, a related tigliane alcohol, have been shown to be present in *Euphorbia cooperi*

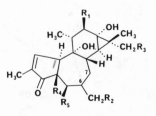

Plant	Alcohol	R_1	R_2	R_3	R_4	R_5	Other
Euphorbia spp., *Baliospermum montanum*	12-Deoxyphorbol	H	OH	H	OH	H	$\Delta^{6,7}$
Euphorbia spp., *B. montanum*	12-Deoxy-16-hydroxyphorbol	H	OH	OH	OH	H	$\Delta^{6,7}$
E. resinifera	12,20-Dideoxyphorbol	H	H	H	OH	H	$\Delta^{6,7}$
Aleurites fordii, *Croton flavens*	16-Hydroxyphorbol	OH	OH	OH	OH	H	$\Delta^{6,7}$
C. flavens	4-Deoxy-16-hydroxyphorbol	OH	OH	OH	H	H	$\Delta^{6,7}$
E. tirucalli	4-Deoxyphorbol	OH	OH	H	H	H	$\Delta^{6,7}$
B. montanum	12-Deoxy-5-hydroxyphorbol	H	OH	H	OH	OH	$\Delta^{6,7}$
Hippomane mancinella, *B. montanum*	12-Deoxy-5-hydroxyphorbol-6α,7α-oxide	H	OH	H	OH	OH	6,7-epoxy

Figure 6.3. Some parent diterpene alcohols of substituted phorbol esters of the Euphorbiaceae.

(Gschwendt and Hecker 1973) and *E. poisonii* (Schmidt and Evans 1977b). Recently, antileukemic and cytotoxic esters of both 12-deoxyphorbol and 12-deoxy-16-hydroxyphorbol were isolated from *Baliospermum montanum* (Ogura et al. 1978) (see figure 6.3).

Further substituted phorbol parent alcohols have been isolated (see figure 6.3). Biologically inactive esters of 12,20-dideoxyphorbol occur in *Euphorbia resinifera* (Hergenhahn, Adolf, and Hecker 1975). The toxic principles of tung fruit (*Aleurites fordii*) have recently been established to be esters of 16-hydroxyphorbol (Okuda et al. 1975). Esters of the same diterpene alcohol, as well of the novel compound 4-deoxy-16-hydroxyphorbol, were recently reported from the roots of *Croton flavens*, a plant that has been implicated as a possible cause of esophageal cancer in Curaçao (Weber and Hecker 1978). The latex of *Euphorbia tirucalli*, when collected in South Africa and Colombia, has been shown to contain complex esters of 4-deoxyphorbol (Fürstenberger and Hecker 1977b; Kinghorn, in press). Mancinellin, one of a series of toxins in *Hippomane mancinella*, the mancineel tree, was reported by Adolf and Hecker (1975a) as the 13-hexadeca-2,4,6-trienoate of 12-deoxy-5-hydroxyphorbol-6α,7α-oxide. A derivative of the same parent alcohol was found in *Baliospermum montanum*, as well as an ester of the novel compound 12-deoxy-5-hydroxyphorbol (Ogura et al. 1978). The possible extraction artifacts, 10β-,Δ^{1-10}-iso- and 4-deoxy-4α-phorbol, obtained in the preparation of phorbol from *Croton tiglium* (Hecker and Schmidt 1974), have been omitted from figure 6.3.

Ingenol Esters

A second category of toxic diterpenoids, ingenol esters, which are based on the hydrocarbon ingenane, also occur in the Euphorbiaceae. The irritant and cocarcinogenic compound ingenol 3-hexadecanoate was found by Hecker and his co-workers to occur in both the herbaceous caper spurge, *Euphorbia lathyris*, and an African succulent member of the genus, *E. ingens* (Hecker 1976). This compound was termed, respectively, L_5 and I_1 upon isolation from these two species, and its structure is shown, along with the parent diterpenoid ingenol, in figure 6.4. The structure of ingenol was established by X-ray crystallography as a triacetate derivative, and it exhibits a three-dimensional structure having similarities to that of phorbol (Zechmeister et al. 1970).

Ingenol was found to be the most common tetracyclic diterpenoid among a representative group of succulent and herbaceous spurges examined by Evans and Kinghorn (1977), and was isolated from 24 species of the genus *Euphorbia*. A fractionation scheme for the isolation of ingenol esters from *E. kansui* has been described by Hirata (1975). Ingenol esters have also been

Figure 6.4. The structure of ingenol and Euphorbia *factor* $I_1 = L_5$.

reported in E. *jolkini* (Uemura and Hirata, 1973), E. *resinifera* (Hergen-hahn, Kusumoto, and Hecker 1974), and E. *serrata* (Upadhyay et al. 1976). In addition, highly unsaturated ingenol esters were recently reported in E. *lathyris* (Adolf and Hecker 1975b) and E. *tirucalli* grown in the Malagasy Republic (Fürstenberger and Hecker 1977a). Esters of ingenol that are par-ticularly interesting are the alkaloidal types in the ornamental crown-of-thorns (*Euphorbia millii*) (Uemura and Hirata 1971, 1977), and ingenol 3,20-dibenzoate in E. *esula*, which has antileukemic activity in mice (Kup-

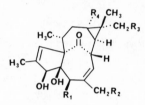

Plant	Alcohol	R_1	R_2	R_3	R_4
Euphorbia spp.	16-Hydroxyingenol	OH	OH	OH	H
Euphorbia kansui	20-Deoxyingenol	OH	H	H	H
E. kansui	13-Oxyingenol	OH	OH	H	OH
E. myrsinites, *E. biglandulosa*	5-Deoxyingenol	H	OH	H	H

Figure 6.5. Parent diterpene alcohols of substituted ingenol esters of the Euphor-biaceae.

chan et al. 1976). Ingenol was reported in a fairly high yield in both the latex and the whole plant of E. *segueiriana* (Upadhyay et al. 1977). Ingenol has so far been isolated in only two species outside the genus *Euphorbia*: in the taxonomically closely related *Elaeophorbia grandiflora* and *El. drupifera* (Kinghorn and Evans 1974).

Analogous to the substituted phorbol alcohols (see figure 6.3), a series of modified ingenol alcohols are shown in figure 6.5. Esters of 16-hydroxy-ingenol occur in *Euphorbia ingens* (Opferkuch and Hecker 1974), E. *lactea* (the ornamental candelabra cactus) (Upadhyay and Hecker 1975), and E. *lathyris* (Adolf and Hecker 1975b). In addition to ingenol, *Euphorbia kansui* has yielded 20-deoxyingenol (Uemura et al. 1975a) and 13-oxyingenol (Ue-mura et al. 1975b). A further substituted ingenol, 5-deoxyingenol, was isolated as a diacetate from *Euphorbia myrsinites* and E. *biglandulosa* (Evans and Kinghorn 1974).

Daphnane Esters

Daphnane esters are a third group of irritant cocarcinogenic toxins in the Euphorbiaceae, and are based on the tricyclic diterpene daphnane. The first compound of this type to be isolated from this family was huratoxin (see figure 6.6), which was found in two notorious toxic plants, *Hura crepitans*

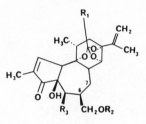

Plant	Compound	R_1	R_2	R_3	Other
Hura crepitans, *Hippomane mancinella*	Huratoxin	$-(CH:CH)_2(CH_2)_8CH_3$	H	OH	6,7-epoxy
H. mancinella	Factor M_2	$-(CH:CH)_3(CH_2)_8CH_3$	H	OH	6,7-epoxy
Excoecaria agallocha	—	$-(CH:CH)_2(CH_2)_4CH_3$	H	OH	6,7-epoxy
Baliospermum montanum	Montanin	$(CH_2)_{10}CH_3$	H	H	6,7-epoxy
Euphorbia resinifera	RL-14	$C_6H_5CH_2$	H	H	$\Delta^{6,7}$
E. resinifera, *E. poisonii,* *E. unispina*	Resiniferatoxin	$C_6H_5CH_2$	$COCH_2$ (with OH, OCH$_3$ substituted phenyl)	H	$\Delta^{6,7}$
E. poisonii	Tinyatoxin	$C_6H_5CH_2$	$COCH_2 \cdot C_6H_4 \cdot pOH$	H	$\Delta^{6,7}$

Figure 6.6. Daphnane esters isolated from the Euphorbiaceae.

(sandbox tree) (Sakata et al. 1971) and *Hippomane mancinella* (mancineel tree) (Adolf and Hecker 1975a). An irritant compound related to huratoxin was also found in the latter species (Adolf and Hecker 1975a). From a completely different species, *Excoecaria agallocha*, a tigliane orthoester with a shorter alkyl chain than that of huratoxin was obtained (Ohigashi et al. 1974). Montanin, from *Baliospermum montanum*, is a related compound with a fully saturated alkyl chain, and is significantly cytotoxic in the P-388 leukemia (PS) system in cell culture (Ogura et al. 1978) (see figure 6.6).

Recent work has demonstrated the presence of a second group of daphnane esters from the Euphorbiaceae that lack 6,7-epoxy and 5β-hydroxy groups (see figure 6.6). Factor RL-14 was obtained by Hecker and coworkers from *Euphorbia resinifera* (Hergenhahn, Adolf, and Hecker 1975). Resinif-

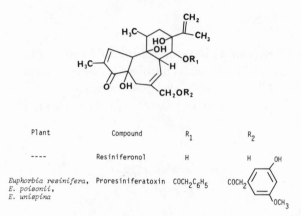

Figure 6.7. A resiniferonol ester isolated from Euphorbia *species.*

eratoxin, a compound with the same benzyl orthoester function as RL-14 and an additional aromatic ester group at C-20, has been isolated from both E. *resinifera* and E. *unispina* (Hergenhahn, Adolf, and Hecker 1975). Subsequent work by Schmidt and Evans (1976) showed resiniferatoxin to be also present in E. *poisonii*, together with a related toxin, tinyatoxin (see figure 6.6).

A final toxin, serving as a possible biosynthetic link between the phorbol group and the daphnane esters was first reported in *Euphorbia resinifera* (Hergenhan, Adolf, and Hecker 1975). Proresiniferatoxin, an ester of resiniferonol, with a benzyl side chain at C-14 (see figure 6.7), upon treatment with traces of trifluoroacetic acid produced resiniferatoxin (Hergenhahn, Adolf, and Hecker 1975). Proresiniferatoxin was later found in E. *poisonii* and E. *unispina* (Schmidt and Evans 1976).

Botany and Ethnobotany of Irritant Cocarcinogenic Euphorbiaceae

Characteristics and Classification

The Euphorbiaceae is considered to be one of the most morphologically variable Angiosperm families, and may be of polyphyletic origin (Webster 1967). The majority of the genera are characterized by unisexual flowers, with highly variable floral characteristics, including normally a superior and three-loculed ovary (Lawrence 1951). Apart from *Euphorbia* (1,600 species) and *Croton* (700 species), 13 other genera of this family have more than 100 species, including *Acalypha* (430), *Antidesma* (160), *Drypetes* (160), *Glochidion* (280), *Jatropha* (150), *Macaranga* (240), *Manihot* (160), *Phyllanthus* (480), and *Tragia* (140) (Lawrence 1951).

The Euphorbiaceae were classified by Pax and Hoffmann (1931) into 283 genera in four subfamilies; Phyllanthoideae, Crotonoideae (including most of the North American genera), and the Australian groups, Porantheroideae and Ricinocarpoideae. Crotonoideae is divided into two tribes, Crotoneae and Euphorbieae. Many irritant species occur in the latter tribe, which is characterized by an inflorescence in the form of a cyathium (Pax and Hoffmann 1931). Each cyathium consists of a single female flower surrounded by several male flowers, and is made up of a surrounding involucre with individual bracts set in it (White, Dyer, and Sloane 1941). Very often the cyathium has real bracts that are colored (Pax and Hoffmann 1931), as in the case of the crown-of-thorns, which are so often mistaken for petals by the lay public. Webster has reviewed past classifications of the Euphorbiaceae and presented the most contemporary classification of the family, based on 5 subfamilies and 52 tribes (Webster 1975).

Economic Botany of Some Toxic Euphorbiaceae

The Euphorbiaceae are of considerable economic importance, and products obtained from this family include castor oil (*Ricinus*), tung oil (*Aleurites*), cassava and tapioca (*Manihot*), and rubber (*Hevea*) (Lawrence 1951). By no means have all of these been shown to contain irritant cocarcinogens, and only the species containing, or suspected of containing, irritant principles will be considered here. Three geographic areas—Nigeria, the United Kingdom and the United States—will be discussed to represent, respectively, examples of native, traditional, and current uses of toxic Euphorbiaceae.

The West African uses of these species have been documented by Irvine (1961), Dalziel (1937), and Kerharo and Adam (1974), and those of Nigerian succulent *Euphorbia* species by Evans and Kinghorn (1975b). Succulent *Euphorbia* species make an excellent natural barrier, in being both spiny and irritant, and they are used for this purpose in villages of the Benue Plateau Province of Nigeria (Evans and Kinghorn 1975b). Native uses of *Euphorbia* species are related to their toxicity, and in general, the less toxic a species, the more ingenious and varied its uses. For example, the highly irritant *Euphorbia poisonii* is used as an arrow poison ingredient and as an insecticide. In contrast, the slightly irritant *E. balsamifera* is used as an oral

Table 6.1 Distribution of Irritant Species of the Euphorbiaceae[a]

Subfamily	Tribe	Genus	References
I. PHYLLANTHOIDEAE Ascherson	3. BRIDELIEAE Muell. Arg.	*Cleistanthus*[b] Hook f. ex Planch	Uphoff 1959; Burkill 1966; von Reis Altschul 1973
	6. SPONDIANTHEAE Webster	*Spondianthus*[b] Engler	Uphoff 1959; Irvine 1961
	7. ANTIDESMEAE (Endl.) Hurusawa	*Antidesma*[b] L.	von Reis Altschul 1973
	10. PHYLLANTHEAE Dumort. Subtribe 10b Flueggeinae Muell. Arg.	*Flueggea*[b] Willd.	Irvine 1961
		Phyllanthus[b] L.	Burkill 1966; Hartwell 1969; von Reis Altschul 1973
II. OLDFIELDIOIDEAE Köhler and Webster			
III. ACALYPHOIDEAE Ascherson	27. PYCNOCOMEAE Hutch. Subtribe 27a Pycnocominae (Hutch.) Webster	*Pycnocoma*[b] Benth.	Irvine 1961
	32. ACALYPHEAE Dumort. Subtribe 32h Claoxylinae Hurusawa	*Mareya*[b] Baill.	*Ibid.*
IV. CROTONOIDEAE Pax	38. MANIHOTEAE (Muell. Arg.) Pax	*Cnidoscolus*[b] Pohl	von Reis Altschul 1973
	42. JOANNESIEAE (Muell. Arg.) Pax, Subtribe 42a Jatrophiinae Meisn.	*Jatropha*[b] L.	Uphoff 1959; Irvine 1961; Kingsbury 1964; Burkill 1966; von Reis Altschul 1973
	Subtribe 42c, Joannesiinae Muell. Arg.	*Joannesia*[b] Vell.	Uphoff 1959

43. CODIAEAE (Pax) Hutch.	*Baliospermum* Bl.	Ogura et al. 1978
	Fontanea [b] Heckel	von Reis Altschul 1973
46. ALEURITIDEAE Hurusawa, Subtribe 46a Aleuritinae (Hurusawa) Webster	*Aleurites* G. Forster	Okuda et al. 1975; Kinghorn, unpublished results
47. CROTONEAE Dumort.	*Croton* L.	Hecker 1977; Weber and Hecker 1978
49. HIPPOMANEAE A. Juss. ex Spach, Subtribe 49c Hippomaninae Griseb.	*Hippomane* L.	Adolf and Hecker 1975a
	Excoecaria L.	Ohigashi et al. 1974
	Grimmeodendron Urb.	Kinghorn, unpublished results
	Sapium P. Br.	Ohigashi et al. 1972
	Sebastiana [b] Spreng.	Uphoff 1959; von Reis Altschul 1973
	Stillingia Gard. ex L.	Kinghorn, unpublished results
51. HUREAE Dumort.	*Hura* L.	Sakata et al. 1971
52. EUPHORBIEAE Dumort., Subtribe 52a Anthosteminae (Kl. and Gcke.) Webster Subtribe 52c Euphorbiinae Hurusawa	*Anthostema* [b] A. Juss.	Irvine 1961
	Euphorbia Dumort.	Hirata 1975; Evans and Kinghorn 1977; Hecker 1977
	Pedilanthus [b] Neck ex. Poit.	Uphoff 1959; Burkill 1966; Hartwell 1969
	Syndenium Boiss.	Kinghorn, unpublished results

V. EUPHORBIOIDEAE Boiss.

[a]Classified according to Webster (1975).
[b]Speculative.

antiseptic, a hemostatic, and a purgative, and in treating gonorrhea (Evans and Kinghorn 1975b).

In the United Kingdom beggars in medieval times used spurges to produce blisters in order to promote pity (Freeman 1971), and their traditional uses as wart cures and mole deterrents are still popular. The use of toxic plants of this family in medicine has declined, although extracts of *Croton tiglium* (drastic purgative) and *Euphorbia resinifera* (external veterinary vesicant) were formerly official drugs (Todd 1967).

Today in the United States many of these toxic plants are ornamental and household plants, including crown-of-thorns (*Euphorbia millii*), pencil tree (*E. tirucalli*), candelabra cactus (*E. lactea*), and coral plant (*Jatropha multifida*) (Saffiotti 1976). The tung oil (*Aleurites fordii*) industry is important for the production of a drying oil with a high refractive index, and meal used as a cattle feed (Kingsbury 1964). An intriguing recent proposal suggested using *Euphorbia lathyris* and *E. tirucalli* as renewable hydrocarbon sources to help rectify the U.S. gasoline shortage (Nielsen et al. 1977).

Distribution of Toxic Diterpenoids at Generic and Specific Levels

Irritant cocarcinogenic diterpenes are currently of great value in studying skin carcinogenesis, and the relationship between their antileukemic and toxic effects warrants further study. Further phytochemical work is necessary to obtain additional examples of these compounds, and the ability to predict as accurately as possible the likelihood of the occurrence of diterpenes in plants of the Euphorbiaceae is desirable.

In an attempt to fulfill this need, table 6.1 presents both the established distribution, obtained from the chemical literature, and the likely distribution, based on ethnobotanical and toxicological literature, of tetracyclic and orthoester diterpenes in genera of the Euphorbiaceae, as arranged by Webster (1975). Genera in which insufficient chemical work has been carried out are designated appropriately in table 6.1, and are included on the basis of reports of at least one species being a drastic purgative, an arrow or fish poison ingredient, an irritant, or a wart cure. The table was compiled by reference to von Reis Altschul (1973); Uphoff (1959); Burkill (1966); Hartwell (1969); Evers and Link (1972); Hardin and Arena (1969); Schmutz, Freeman, and Reed (1968); Lampe and Fagerström (1968); Irvine (1961); Dalziel (1937); Kerharo and Adam (1974); and Kingsbury (1964).

Table 6.2 Distribution of Irritant Diterpenoids in the Genus *Euphorbia*[a]

Section	Subsection	Species Investigated	Irritant Species	Diterpenoids Present
I. *Anisophyllum* (Haw.) Roep.	*Hypericifoliae* Boiss.	1	0	none
III. *Poinsettia* (Grah.) Boiss.	—	2	0	none
VII. *Euphorbium* Benth.	*Tirucalli* Boiss.	4	2	4-deoxyphorbol; unknown
	Diacanthium Boiss.	21	21	ingenol; 12-deoxyphorbol; phorbol; 12-deoxy-16-hydroxyphorbol; resiniferatoxin
IX. *Tithymalus* (Hall.) Boiss.	*Decussatae* Boiss.	1	1	ingenol
	Pachycladae Boiss.	1	1	12-deoxyphorbol
	Galarrhaei Boiss.	8	8	ingenol; 12-deoxyphorbol; 5-deoxyingenol; unknown
	Esulae Boiss.	11	10	ingenol; 5-deoxyingenol; unknown
	Myrsiniteae Boiss.	2	2	ingenol; 5-deoxyingenol

[a]Classified according to Pax and Hoffmann (1931).

Table 6.1 shows that all of the toxic diterpenes that have been discovered to date occur in genera of a fairly narrow series of tribes in the subfamilies Crotonoideae and Euphorbiodeae. Species of the genera in Webster's subtribes Hippomaninae and Euphorbiinae are particularly prominent in yielding these compounds. Although none of these compounds has yet been discovered in the subfamily Phyllanthoideae, the several references to the toxic action of certain species are worthy of investigation.

Thus far one study has attempted to correlate the distribution of irritant diterpenes within morphological groups of a genus of the Euphorbiaceae. Evans and Kinghorn (1977) subjected about 60 members of the genus *Euphorbia* to a phytochemical screen, and table 6.2 is a summary of their results. The identification of parent diterpene alcohols was accomplished using chromatographic and spectroscopic methods after hydrolyzing the irritant esters and forming stable acetates in each extract. For every species shown to contain a tetracyclic or orthoester diterpene, a positive irritancy test on the mouse ear inflammation model was observed (Kinghorn and Evans 1975b). The number of nonirritant species was surprisingly low (about 10 percent), and all but one fell into the taxonomic sections *Anisophyllum, Poinsettia,* and *Euphorbium* subsection *Tirucalli* of the Pax and Hoffmann 1931 classification. One interesting result was the lack of either irritant activity or tetracyclic diterpenes in samples of *E. pulcherrima* (poinsettia), from two geographic origins, thereby confirming the lack of toxicity observed in rats (Stone and Collins 1971). This result, coupled with the unusual triterpene profile of *E. pulcherrima* and other species in the section *Poinsettia* (Ponsinet and Ourisson 1968), lends support to the suggestion of a botanical dissociation of this section from the main corpus of *Euphorbia* species (Webster 1967).

Toxicological Activity of Diterpene Esters of the Euphorbiaceae

Action as Tumor-Promoters in Two-Stage Carcinogenesis Experiments

Several reviewers have described the events leading up to Berenblum's observation of the potent tumor-promoting (cocarcinogenic) activity of croton oil, and the subsequent use of croton oil and phorbol esters in "two-stage" experiments on mouse skin (Boutwell 1974; Berenblum 1975; Hecker 1976). Such studies, also known as Berenblum experiments, involve two distinct stages in the generation of a neoplastic cell. The application of a subcarcinogenic dose of a carcinogen "initiates" irreversibly a "latent tumor cell" that is macroscopically indistinct from normal epidermal cells. This cell can then be reversibly "promoted" to form a neoplastic cell by subsequent applications of either an adequate dose of the carcinogen or repeated doses of a tumor-promoter such as croton oil or its diterpenoid constituents (Boutwell 1974; Hecker 1976). Repeated applications of a tumor-promoter to a normal epidermal cell lead to hyperplasia but not malignancy (Berenblum and Shubik 1947; Boutwell 1974). The present conception of the biological properties of tumor-promoters has been reported by Weinstein and Troll (1977). Chemical carcinogenesis in mouse skin is now considered to be more than merely a two-step process, and up to five or six discrete events have been suggested by some authorities (Marx 1978).

The significance of the phenomenon of tumor-promotion has increasingly been realized as promotion has been observed in tissues and organisms other than the skin of mice (Marx 1978). Phorbol, an inactive cocarcinogen when administered topically on mouse skin (Berenblum 1975), when given systemically after a subeffective dose of various carcinogens has been shown to potentiate not only mouse skin carcinogenesis (Armuth and Berenblum 1976) but also mouse lung and liver carcinogenesis (Armuth and Berenblum 1972), rat mammary carcinogenesis (Armuth and Berenblum 1974), and leukemias in mice (Berenblum and Lonai 1970).

The possibility not only of devising rapid screening procedures for tumor-promoters but also of developing *in vitro* methods for their biochemical study was first realized by Lasne, Gentil, and Chouroulinkov (1974), when they demonstrated two-stage carcinogenesis in tissue culture. Malignant transformation of cultured rat embryonic fibroblasts was obtained after initiation with benzo[a]pyrene and promotion by TPA (Lasne, Gentil, and Chouroulinkov 1974).

Structural Effects of Diterpene Esters on Toxicological Activity

Several biological test methods were developed by the Hecker group to monitor the fractionation of the active principles of croton oil (Hecker and

Schmidt 1974). More refined methods of assessing the toxicity of the diterpene esters of the Euphorbiaceae (the parent alcohols being inactive), which were capable of being evaluated statistically, were also reported (Hecker 1968). Quantitative Berenblum experiments on mouse skin, employing an initial subeffective carcinogen application and often over 30 weeks of repeated promoter applications, have been used to assess cocarcinogenic potency (Hecker 1968). Irritant dose 50 percent (ID_{50}) is a quantitative expression of irritancy on the mouse ear, where the results are obtained up to 24 hours after the application of test solutions (Kinghorn and Evans 1975b).

Accounts of the cocarcinogenic potencies of a range of diterpene esters have appeared in the literature (Hecker 1971; Fürstenberger and Hecker 1972). TPA (see figure 6.1), the most potent tumor promoter from croton oil (Hecker and Schmidt 1974) and of several semisynthetic phorbol esters (Boutwell 1974; Berenblum 1975) was significantly more active than ingenol hexadecanoate (see figure 6.4) (Fürstenberger and Hecker 1972). The semisynthetic compound phorbol 12,13-didecanoate (Fürstenberger and Hecker 1972) and, in addition, the recently described esters of 16-hydroxyphorbol and 4-deoxy-16-hydroxyphorbol (Weber and Hecker, 1978) exhibit a similar cocarcinogenic potency to TPA.

A summary of the structural requirements for irritancy and modifications of structure leading to an increase in irritancy in a group of 12-deoxyphorbol esters is shown in figure 6.8. Conclusions were reached on the basis of the determination of ID_{50}'s of compounds isolated from various *Euphorbia* species (Evans, Kinghorn, and Schmidt 1975). Other studies that have shown highly irritant compounds are those on 12-deoxyphorbol-13-tetradecanoate (Gschwendt and Hecker 1974) and resiniferatoxin, which is particularly active after 2–4 hours (Hergenhahn, Adolf, and Hecker 1975).

Initial work demonstrated that the phorbol diester croton oil factors had both irritant and cocarcinogenic activities (Hecker 1971). However, qualitative differences became apparent, as exemplified by the fact that TPA (see figure 6.1) is the most cocarcinogenic, though not the most irritant, croton oil toxic factor (Hecker and Schmidt 1974). Recent observations of highly irritant noncocarcinogenic semisynthetic phorbol esters and 4-deoxyphorbol esters in *Euphorbia·tirucalli* indicate independent structural requirements for the irritant and cocarcinogenic activities of this class of compounds (Fürstenberger and Hecker 1972).

Two other studies have used different animal models to investigate structure–activity relationships of the diterpene esters of the Euphorbiaceae. The

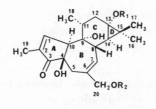

Requirements for Irritancy:

1. A rigid tetracyclic structure

2. Ar. A/B *trans*-junction

3. Ring C in the *cis*-configuration

4. A 4-β function

5. An ester function at C-13

Irritancy Increases:

1. In 12-deoxyphorbol. 13-monoesters (R_1 = acyl; R_2 = H)
   compared with 20-O-acyl diester analogs (R_1, R_2 = acyl)

2. With chain-length increase in R_1

3. With the introduction of unsaturation in R_1

Figure 6.8. The structure–irritancy relationship of 12-deoxyphorbol esters.

piscicidal activity of tigliane orthoesters was found to increase with acyl chain length when tested on killifish (Sakata, Kawazu, and Mitsui 1971; Ohigashi et al. 1974). Phorbol esters were shown to have a parallel toxicity on brine shrimp larvae (Kinghorn, Harjes, and Doorenbos 1977) to their tumor-promoting potency (Boutwell 1974).

Cocarcinogenic Irritant Euphorbiaceae as a Potential Health Hazard

In the United States the members of the spurge family creating a health hazard to humans are mainly ornamental species (Saffiotti 1976); those hazardous to livestock are indigenous species (Kingsbury 1964). Irritant species of the Euphorbiaceae produce lesions of the skin in both man and livestock that are liable to secondary infection (Evans and Kinghorn 1975b). In humans, skin inflammation symptoms can last up to three weeks (Kingsbury 1964). Direct contact of irritant latex to the eyes, or transference of the irritant principles to the eyes via the fingers, can lead to severe keratoconjunctivitis (Lampe and Fagerström 1968) and even to temporary blindness (Kingsbury 1964). The herbaceous spurges common to the United States are as irritant to the mouse ear as the ornamental *Euphorbia* species (Kinghorn and Evans 1975b).

The problem and symptoms of human and livestock poisoning by ingestion of plant parts of the irritant Euphorbiaceae have been surveyed previously (Kingsbury 1964; Schmutz, Freeman, and Reed 1968; Hardin and Arena 1969). In humans, internal consumption is probably greatest in species with a prominent seed, such as *Aleurites*, although a case has been documented in which five women used the fruits of *Euphorbia lathyris* (caper spurge) in their cooking in place of capers (Kingsbury 1964).

Symptomatic approaches for the treatment of toxic reactions in humans by members of the Euphorbiaceae have been proposed (Todd 1967; Lampe and Fagerström, 1968; Hardin and Arena 1969).

Much interest has been aroused by the possibility that plants are causative agents in the genesis of human cancer (Saffiotti 1976). Farnsworth and his colleagues (1976) have extensively reviewed the literature on plant oncogens. In the case of household plants of the spurge family, it has been suggested that the cocarcinogenic principles may be inhaled as a spray when, for example, a branch is broken (Saffiotti 1976). However, the danger of carcinogenesis, even if trace amounts of carcinogens are omnipresent, is unlikely, since repeated applications of the spurge tumor-promoting principles would be necessary to elicit a cancer (Saffiotti 1976). The same point was stressed by Falk (1976) in a discussion on the effect of cocarcinogens in the development of human cancer.

Even though the cocarcinogenic effect of these toxins has not been demonstrated in humans, knowledge of the extent of human carcinogens in the environment and of combinative effects of cocarcinogens in both animals and humans is rudimentary. Clearly, a great deal of work is necessary before the whole extent of the public health hazard posed by these diterpene esters of the Euphorbiaceae is fully realized.

REFERENCES

Adolf, W. and E. Hecker. 1975a. "On the Irritant and Cocarcinogenic Principles of *Hippomane mancinella.*" *Tetrahedron Lett.* 1975:1587–90.

—— 1975b. "On the Active Principles of the Spurge Family. 3. Skin Irritant and Cocarcinogenic Factors of the Caper Spurge." *Z. Krebsforsch. Klin. Oncol.* 84:325–44.

Armuth, V. and I. Berenblum. 1972. "Systemic Promoting Action of Phorbol in Lung and Liver Carcinogenesis in AKR Mice." *Cancer Res.* 32:2250–62.

—— 1974. "Promotion of Mammary Carcinogenesis and Leukemogenic Action by Phorbol in Virgin Female Wistar Rats." *Cancer Res.* 34:2704–7.

—— 1976. "Phorbol as a Possible Systemic Promoting Agent for Skin Carcinogenesis." *Z. Krebsforsch. Klin. Oncol.* 85:79–82.

Berenblum, I. 1941. "The Mechanism of Carcinogenesis. A Study of the Significance of Cocarcinogenic Action and Related Phenomena." *Cancer Res.* 1:807–14.

Berenblum, I. and P. Shubik. 1947. "A New Quantitative Approach to the Study of the Stages of Chemical Carcinogenesis in the Mouse Skin." *Br. J. Can.* 1:383–390.

Berenblum, I. and V. Lonai, 1970. "Leukemogenic Action of Phorbol." *Cancer Res.* 30:2744–48.

Berenblum, I. 1975. "Sequential Aspect of Chemical Carcinogenesis: Skin." In F. Becker, ed., *Cancer Etiology*, vol. 1, *Chemical Carcinogenesis*, pp. 323–44. New York: Plenum Press.

Boutwell, R. 1974. "The Function and Mechanisms of Promoters of Carcinogenesis." *C.R.C. Crit. Rev. Toxicol.* 2:419–43.

Burkill, I. 1966. *A Dictionary of the Economic Plants of the Malay Peninsula*, vols. 1, 2. Kuala Lumpur: Ministry of Agriculture and Cooperative.

Dalziel, J. 1937. *The Useful Plants of West Tropical Africa*. London: Crown Agents.

Evans, F. and A. Kinghorn. 1974. "A New Ingenol Type Diterpene from the Irritant Fractions of *Euphorbia myrsinites* and *E. biglandulosa.*" *Phytochemistry* 13:2324–25.

—— 1975a. "New Diesters of 12-Deoxyphorbol." *Phytochemistry* 14:1669–70.

—— 1975b. "The Succulent Euphorbias of Nigeria, Part 1." *Lloydia* 38:363–65.

Evans, F., A. Kinghorn, and R. Schmidt. 1975. "Some Naturally Occurring Skin Irritants." *Acta Pharmacol. Toxicol.* 37:250–56.

Evans, F. and R. Schmidt. 1976. "Two New Toxins from the Latex of *Euphorbia poisonii.*" *Phytochemistry* 15:333–35.

Evans, F. 1977. "A New Phorbol Triester from the Latices of *Euphorbia frankiana* and *E. coerulescens.*" *Phytochemistry* 16:395–96.

Evans, F. and A. Kinghorn. 1977. "A Comparative Phytochemical Study of the Diterpenes of Some Species of the Genera *Euphorbia* and *Elaeophorbia* (Euphorbiaceae)." *J. Linn. Soc. London. Bot.* 74:23–35.

Evers, R. and R. Link. 1972. *Poisonous Plants of the Midwest and Their Effect on Livestock*, Special Publication 24. Champaign: University of Illinois, College of Agriculture.

Falk, H. 1976. "Possible Mechanisms of Combination Effects in Chemical Carcinogenesis." *Oncology* 33:77–85.

Farnsworth, N., A. Bingel, H. Fong, A. Saleh, G. Christenson, and S. Saufferer. 1976. "Oncogenic and Tumor-Promoting Spermatophytes and Pteridophytes and Their Active Principles." *Cancer Treat. Rep.* 60:1171–1214.

Freeman, M. 1971. *Herbs for the Medieval Household, for Cooking, Healing and Divers Uses.* New York: Metropolitan Museum of Art.

Fürstenberger, G. and Hecker, E. 1972. "Zum Wirkungsmechanismus cocarcinogener Pflanzeninhaltsstoffe." *Planta Med.* 22:241–266.

—— 1977a. "New Highly Irritant Euphorbia Factors from the Latex of *Euphorbia tirucalli* L." *Experientia* 33:986–88.

—— 1977b. "The New Diterpene 4-Deoxyphorbol and Its Highly Unsaturated Irritant Esters." *Tetrahedron Lett.* 1977:925–28.

Gschwendt, M. and E. Hecker. 1973. "Über die Wirkstoffe der Euphorbiaceen, I. Hautreizende und cocarcinogene Faktoren aus *Euphorbia cooperi* N.E. Br." *Z. Krebsforsch. Klin. Oncol.* 80:335–50.

—— 1974. "Über die Wirkstoffe der Euphorbiaceen. 2. Hautreizende und cocarcinogene Faktoren aus *Euphorbia triangularis* Desf." *Z. Krebsforsch. Klin. Oncol.* 81:193–210.

Hardin, J. and J. Arena. 1969. *Human Poisoning from Native and Cultivated Plants.* Durham, N.C.: Duke University Press.

Hartwell, J. 1969. "Plants Used Against Cancer. A Survey." *Lloydia* 32:153–216.

Hecker, E., H. Bresch and I. Meyer. 1965. "Cocarcinogenic Ingredient of Croton Oil." *Fette, Seifen, Anstrichmittel* 67:78–81.

Hecker, E. 1968. "Cocarcinogenic Principles from the Seed Oil of *Croton tiglium* and from Other Euphorbiaceae." *Cancer Res.* 28:2338–49.

—— 1971. "Cocarcinogens from Euphorbiaceae and Thymeleaceae." In H. Wagner and L. Hörhammer, eds., *Pharmacognosy and Phytochemistry,* pp. 147–65. Berlin, Heidelberg, New York: Springer-Verlag.

Hecker, E. and R. Schmidt. 1974. "Phorbolesters—The Irritants and Cocarcinogens of *Croton tiglium* L." *Fortsch. Chem. Org. Naturst.* 31:377–467.

Hecker, E. 1976. "Aspects of Cocarcinogenesis." In T. Symington and R. Carter, eds., *Scientific Foundations of Oncology,* pp. 310–18. London: Heinemann.

—— 1977. "New Toxic, Irritant and Cocarcinogenic Diterpene Esters from Euphorbiaceae and from Thymelaeaceae." *Pure Appl. Chem.* 49:1423–31.

Hergenhahn, M., S. Kusumoto, and E. Hecker. 1974. "Diterpene Esters from 'Euphorbium' and Their Irritant and Cocarcinogenic Activity." *Experientia* 30:1438–40.

Hergenhahn, M., W. Adolf, and E. Hecker. 1975. "Resiniferatoxin and Other Esters of Novel Polyfunctional Diterpenes from *Euphorbia resinifera* and *unispina.*" *Tetrahedron Lett.* 1975:1595–98.

Hirata, Y. 1975. "Toxic Substances of the Euphorbiaceae." *Pure Appl. Chem.* 41:175–99.

Hoppe, W., F. Brandl, I. Strell, M. Röhrl, I. Gassmann, E. Hecker, H. Bartsch, G. Kreibich, and C. Szczepanski. 1967. "X-Ray Structure Analysis of Neophorbol." *Angew. Chem. Int. Edit.* 6:809–10.

Irvine, F. 1961. *Woody Plants of Ghana.* London: Oxford University Press.

Kerharo, J. and J. Adam. 1974. *La Pharmacopée Sénégalaise Traditionelle.* Paris: Vergot Frères.

Kinghorn, A. and F. Evans. 1974. "Occurrence of Ingenol in *Elaeophorbia* Species." *Planta Med.* 26:150–54.

—— 1975a. "Skin Irritants of *Euphorbia fortissima.*" *J. Pharm. Pharmacol.* 27:329–33.

—— 1975b. "A Biological Screen of Selected Species of the Genus *Euphorbia* for Skin Irritant Effects." *Planta Med.* 28:325–35.

Kinghorn, A., K. Harjes, and N. Doorenbos. 1977. "Screening Procedure for Phorbol Esters Using Brine Shrimp (*Artemia salina*) Larvae." *J. Pharm. Sci.* 66:1369–70.

Kinghorn, A. In press. "Characterization of an Irritant 4-Deoxyphorbol Diester from *Euphorbia tirucalli* L." *J. Nat. Prod.*

—— Unpublished results.

Kingsbury, J. 1964. *Poisonous Plants of the United States and Canada.* Englewood Cliffs, N.J.: Prentice-Hall.

Kupchan, S., I. Uchida, A. Branfman, R. Dailey, Jr., and B. Yu Fei. 1976. "Antileukemic Principles Isolated from Euphorbiaceae Plants." *Science* 191:571–72.

Lampe, K. and R. Fagerström. 1968. *Plant Toxicity and Dermatitis.* Baltimore: Williams and Wilkins.

Lasne, C., A. Gentil, and I. Chouroulinkov. 1974. "Two-Stage Malignant Transformation of Rat Fibroblasts in Cell Culture." *Nature (London)* 247:490–91.

Lawrence, G. 1951. *Taxonomy of Vascular Plants,* New York: Macmillan.

Lewis, W. and M. Elvin-Lewis. 1977. *Medical Botany.* New York: Wiley.

Marx, J. 1978. "Tumor Promoters: Carcinogenesis Gets More Complicated." *Science* 201:515–18.

Nielsen, P., H. Nishimura, J. Otvos, and M. Calvin. 1977. "Plant Crops as a Source of Fuel and Hydrocarbon-like Materials." *Science* 198:942–44.

Ogura, M., K. Koike, G. Cordell, and N. Farnsworth. 1978. "Potential Anticancer Agents 8. Constituents of *Baliospermum montanum* (Euphorbiaceae)." *Planta Med.* 33:128–43.

Ohigashi, H., K. Kawazu, K. Koshimizu, and T. Mitsui. 1972. "A Piscicidal Constituent of *Sapium japonicum.*" *Agric. Biol. Chem.* 36:2529–37.

Ohigashi, H., H. Katsumata, K. Kawazu, K. Koshimizu, and T. Mitsui. 1974. "A Piscicidal Constituent of *Excoecaria agallocha.*" *Agric. Biol. Chem.* 38:1093–95.

Okuda, T., T. Yoshida, S. Koike, and N. Toh. 1975. "New Diterpene Esters from *Aleurites fordii* Fruits." *Phytochemistry* 14:509–15.

Opferkuch, H. and E. Hecker. 1974. "New Diterpene Irritants from *Euphorbia ingens.*" *Tetrahedron Lett.* 1974:261–64.

Pax, F. and K. Hoffmann. 1931. "Euphorbiaceae." In A. Engler and K. Prantl, eds., *Die Natürlichen Pflanzenfamilien,* 2nd ed., 19C. Leipzig: Engelmann.

Pettersen, R., G. Ferguson, L. Crombie, M. Games, and D. Pointer. 1967. "The Structure and Stereochemistry of Phorbol, Diterpene Parent of Co-carcinogens of Croton Oil." *Chem. Commun.* 1967:716–17.

Ponsinet, G. and G. Ourisson. 1968. "Études chimiotaxonomiques dans la famille des Euphorbiacées. 3. Repartition des triterpènes dans les latex d'*Euphorbia.*" *Phytochemistry* 7:89–98.

Saffiotti, U. 1976. "Are Some House and Garden Plants Carcinogenic?" *J. Am. Med. Assoc.* 235:2344.

Sakata, K., K. Kawazu, and T. Mitsui. 1971. "Studies on a Pisicicidal Constituent of *Hura crepitans.* Part 1. Isolation and Characterization of Huratoxin and Its Piscicidal Activity." *Agric. Biol. Chem.* 35:1084–91.

Sakata, K., K. Kawazu, T. Mitsui, and N. Masaki. 1971. "Structure and Stereochemistry of Huratoxin, a Piscicidal Constituent of *Hura crepitans.*" *Tetrahedron Lett.* 1971:1141–44.

Schmidt, R. and F. Evans. 1976. "A New Aromatic Ester Diterpene from *Euphorbia poisonii.*" *Phytochemistry* 15:1778–79.

—— 1977a. "The Succulent *Euphorbias* of Nigeria. 2. Aliphatic Diterpene Esters of the Latices of *E. poisonii* Pax and *E. unispina* N.E. Br." *Lloydia* 40:225–29.

—— 1977b. "Candletoxins A and B, 2 New Aromatic Esters of 12-Deoxy-16-Hydroxyphorbol, from the Irritant Latex of *Euphorbia poisonii* Pax." *Experientia* 33:1197–98.

Schmutz, E., B. Freeman, and R. Reed. 1968. *Livestock-Poisoning Plants of Arizona.* Tucson: University of Arizona Press.

Stone, R. and W. Collins. 1971. "*Euphorbia pulcherrima:* Toxicity to Rats." *Toxicon* 9:301–2.

Todd, R., ed. 1967. *Martindale, Extra Pharmacopoeia,* 25th ed. London: Pharmaceutical Press.

Uemura, D. and Y. Hirata. 1971. "The Isolation and Structures of Two New Alkaloids, Milliamines A and B, Obtained from *Euphorbia millii.*" *Tetrahedron Lett.* 1971:3673–76.

—— 1973. "Isolation and Structures of Irritant Substances Obtained from *Euphorbia* Species (Euphorbiaceae)." *Tetrahedron Lett.* 1973:881–84.

Uemura, D., H. Ohwaki, Y. Hirata, Y-P. Chen, and H-Y. Hsu. 1975a. "Isolation and Structures of 20-Deoxyingenol, New Diterpene, Derivatives and Ingenol Derivatives Obtained from 'Kansui.' " *Tetrahedron Lett.* 1975:2527–28.

Uemura, D., Y. Hirata, Y-P. Chen, and H-Y. Hsu. 1975b. "New Diterpene, 13-Oxyingenol, Derivative isolated from *Euphorbia kansui* Liou." *Tetrahedron Lett.* 1975:2529–32.

Uemura, D. and Y. Hirata. 1977. "The Isolation and Structure of Toxic Principles, Milliamines A, B, and C, from *Euphorbia millii.*" *Bull. Chem. Soc. Japan* 50:2005–9.

Upadhyay, R. and E. Hecker. 1975. "Diterpene Esters of the Irritant and Cocarcinogenic Latex of *Euphorbia lactea.*" *Phytochemistry* 14:2514–15.

—— 1976. "A New Cryptic Irritant and Cocarcinogen from the Seeds of *Croton sparciflorus.*" *Phytochemistry* 15:1070–72.

Upadhyay, R., M. Ansarin, M. Zarintan, and P. Shakui. 1976. "Tumor Promoting Constituent of *Euphorbia serrata* L. Latex." *Experientia* 32:1196–97.

Upadhyay, R., K. Khalesi, G. Kharazi, and M. Ghaisarzadeh. 1977. "Isolation of Ingenol from the Plants of Euphorbiaceae." *Ind. J. Chem.* 15B:294.

Uphoff, J. 1959. *Dictionary of Economic Plants.* Weinheim: Engelmann.

Van Duuren, B. and L. Orris. 1965. "Tumor-Enhancing Principles of *Croton tiglium.*" *Cancer Res.* 25:1871–75.

Von Reis Altschul, S. 1973. *Drugs and Foods from Little Known Plants. Notes in Harvard University Herbaria.* Cambridge, Mass.: Harvard University Press.

Weber, J. and E. Hecker. 1978. "Cocarcinogens of the Diester Type from *Croton flavens* L. and Esophageal Cancer in Curaçao." *Experientia* 34:679–682.

Webster, G. 1967. "The Genera of Euphorbiaceae in the Southeastern United States." *J. Arnold Arb. Harv. Univ.* 48:303–430.

—— 1975. "Conspectus of a New Classification of the Euphorbiaceae." *Taxon* 24:593–601.

Weinstein, I. and W. Troll. 1977. "National Cancer Institute Workshop on Tumor Promotion and Cofactors in Carcinogenesis" (meeting report). *Cancer Res.* 37:3461–63.

White, A., R. Dyer, and B. Sloane. 1941. *The Succulent Euphorbieae (Southern Africa)*; vols. 1, 2. Pasadena, Calif.: Abbey Gardens.

Zechmeister, K., F. Brandl, W. Hoppe, E. Hecker, H. Opferkuch, and W. Adolf. 1970. "Structure Determination of the new Tetracyclic Diterpene Ingenol-Triacetate with Triple Product Methods." *Tetrahedron Lett.* 1970:4075–78.

7   The Poisonous Anacardiaceae

Harold Baer

The Anacardiaceae do not contain toxic substances in the usual sense of "a liquid, solid or gaseous substance which has an inherent property that tends to destroy life or impair health" (Stein 1967). Normally, the toxic properties are exhibited not at the first exposure but only after additional exposures. Certain plants of this family induce an immunologic disease that requires a first contact to sensitize the individual. When a sensitized individual comes into contact with an offending plant, the major effect is a dermatitis that usually starts two to five days following exposure and is characterized by an area of redness or erythema and vesicles containing a clear fluid. Since 40 to 75 percent of the U.S. population is sensitive to some degree (see figure 7.1), the disease regularly afflicts millions of individuals (Kligman 1958; Epstein 1959; Epstein et al. 1974). Because of the widespread occurrence of this condition, it is also an occupational problem (California Dept. of Health 1973; Epstein 1974). The fact that not everyone is sensitive to poisonous Anacardiaceae is partly due to lack of exposure to the plants. However, there appears to be a segment of the population that does not get this disease, regardless of exposure, a phenomenon that has been inadequately explored.

Botany of the Anacardiaceae

The family Anacardiaceae contains about sixty genera. These include the cashew nut tree (*Anacardium occidentale*), the mango tree (*Mangifera indica*), the poison wood tree (*Metopium toxiferum*) found in Florida and parts of the West Indies, and *Toxicodendron* species found in North America and Eastern Asia. In the United States the most important and widespread members of the genus *Toxicodendron* are poison ivy (*T. radicans*, which exists as a number of subspecies) and western poison oak (*T. diversilobum*).

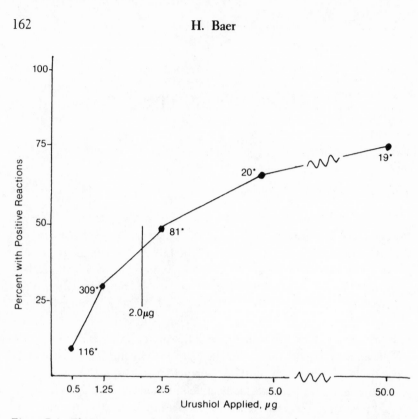

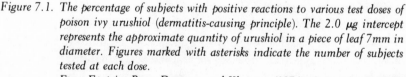

Figure 7.1. The percentage of subjects with positive reactions to various test doses of poison ivy urushiol (dermatitis-causing principle). The 2.0 µg intercept represents the approximate quantity of urushiol in a piece of leaf 7mm in diameter. Figures marked with asterisks indicate the number of subjects tested at each dose.
From Epstein, Baer, Dawson, and Khurana (1974). Copyright © 1977, American Medical Association.

There is also an eastern poison oak (*T. toxicarium*), of limited distribution in the Atlantic and Gulf coastal plains, and poison sumac (*T. vernix*). The distribution of some of these plants in the United States has been described in detail by Gillis (1971).

The literature contains considerable controversy over the nomenclature of the genus *Toxicodendron*. Most of the earlier literature usually refers to poison ivy and poison oak as being in the genus *Rhus*. In addition, the medical literature normally refers to the toxic effects of these plants as Rhus dermatitis. The most recent taxonomic study places the toxic plants of the Anacardiaceae in the genus *Toxicodendron* while the genus *Rhus* contains nontoxic plants (Gillis 1971).

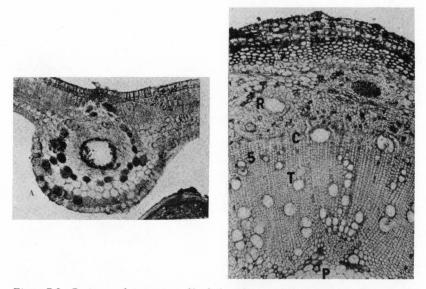

Figure 7.2. Resin canals in section of leaf rib and stem of the Japanese lacquer tree. R, resin canal; T, tracheidal tube; C, cambium; P, pith.
From McNair (1923).

Certain plants of the Anacardiaceae—for example, *Toxicodendron verniciferum* (formerly *Rhus vernicifera*), the lacquer tree of Japan—are of great economic importance. The sap of this plant, which contains a dermatitis-producing substance, oxidizes and polymerizes to a fine black lacquer. The sap flows in resin canals (see figure 7.2), which are in the roots, stem, leaves and immature fruit. Resin canals are absent in the anthers, pollen, trichomes, epidermis, cork cells, and xylem of this plant. Since the sap is only in canals, the undamaged surfaces of the plant, including the leaves and stems, are harmless. However, any small damage due to insect attack or mechanical trauma that permits sap flow onto a leaf surface renders that part capable of inducing dermatitis. The botany of *Toxicodendron* has been extensively reviewed by McNair (1923) and Gillis (1971).

Chemistry of the Toxic Principles

The active principles of only a few of the Anacardiaceae have been studied in detail. However, in all of the plants studied, the dermatitis-causing substances are insoluble in water but soluble in such solvents as alcohol and

Table 7.1 Composition of Urushiol from Various Sources

	C₁₅ Catechols				C₁₇ Catechols				
	Percent Side Chain								
	Saturated	Monoene	Diene	Triene	Saturated	Monoene	Diene	Triene	Tetraene
Poison ivy									
C-1[a]	0	25.8	67.8	3.3	0.2	0.7	1.8	0.4	0
C-2[a]	0	41.6	50.3	3.4	0	0.8	3.1	0.8	0
C-3[a]	6.0	13.5	63.0	16.5	0	0.3	0.7	0	0
M-1[b]	0	15.4	17.9	62.5	0	0	0	4.2	0
F-1[c]	7.6	25.2	38.8	4.3	0	0.8	22.0	1.3	0
F-2[c]	0	3.1	83.1	9.4	0	0.4	2.6	1.4	0
F-3[c]	6.0	11.6	60.0	12.5	0	2.1	6.7	1.1	0
Poison oak									
C-1[d]	22.3	8.4	2.7	0	2.1	10.7	18.7	35.1	0
M-1[e]	3.4	1.4	0.2	0.6	0	5.7	24.9	62.5	1.3
Poison sumac									
C-1[f]	15.3	41.4	32.7	10.6	0	0	0	0	0
Poison wood									
C-1[g]	3.9	48.6	41.4	0	0	1.8	3.3	1.0	0

[a]Collected at Pearl River, N.Y.
[b]Collected at Oxford, Miss.
[c]Collected at Bethesda, Md.
[d]Collected in northern California; extract prepared at Columbia University, New York, N.Y.
[e]Collected in northern California; extract prepared at University, Miss.
[f]Collected in northern New Jersey.
[g]Collected in the Bahamas.

acetone. Thus far, four basic types of substances have been characterized: pentadecylresorcinols from cashew nut oil (see figure 7.3, I); pentadecylcatechols from poison ivy, poison wood tree, poison sumac, and Japanese lacquer (see figure 7.3, II); heptadecylcatechols from western poison oak (see figure 7.3, III); and pentadecylphenol from cashew nut oil (see figure 7.3, IV). The latter substance is either a nonsensitizer or, at best, a very weak

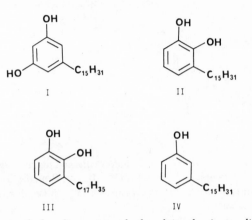

Figure 7.3. Classes of phenolic compounds found in the Anacardiaceae. I, Pentadecylresorcinol; II, Pentadecylcatechol; III, Heptadecylcatechol; IV, Pentadecylphenol.

one. The basic chemistry of cashew nut oil substances has been described (Dawson 1956) and will not be detailed in this review because of its limited interest. Curiously, similar types of substances are found in the fruit of the taxonomically unrelated gingko tree, which may produce a dermatitis from contact (Becker and Skipworth 1975).

The active principle (urushiol) from the various species of *Toxicodendron* is not a single substance but, rather, a group of closely related substances varying by the number and position of double bonds in the side chain. The major components of poison ivy urushiol are shown in figure 7.4. Although the toxic constituents of poison ivy are mainly pentadecylcatechols, there are also some heptadecylcatechols present. In contrast, poison oak contains mainly heptadecylcatechols, with some pentadecylcatechols. The results of analysis of a number of samples of poison ivy, poison oak, poison sumac, and poison wood tree are given in table 7.1 (Gross, Baer and Fales 1975; Corbett and Billets 1975).

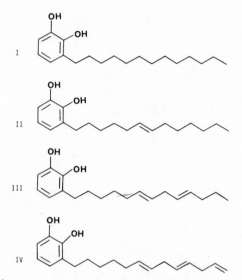

Figure 7.4. *The four major types of pentadecylcatechols found in poison ivy urushiol. I, 3-n-Pentadecylcatechol (PDC); II, 3-n-Pentadec-8'-enyl-catechol; III, 3-n-Pentadec-8',11'-dienyl-catechol; IV, 3-n-Pentadec-8',11',14'-trienyl-catechol.*

Immunological Studies

Alkylcatechols, when administered topically or orally, are rapidly absorbed and become tissuebound, and are excreted in the urine. Their rapid absorption rate has been convincingly demonstrated in experiments on guinea pigs; when the site of application is excised within an hour, the animals still become sensitized (Godfrey and Baer 1971). Since the nonsensitizer pentadecylveratrole is absorbed and distributed in a similar manner to the chemically related sensitizer pentadecylcatechol (PDC), there appears to be no direct correlation between the absorption rate and sensitizing capacity of these compounds (Godfrey, Baer, and Watkins 1971). Dermal application of alkylcatechols or, more effectively, injection in Freund's adjuvant results in delayed contact sensivity. This occurs not only in man but also in guinea pigs, goats, and sheep; rhesus monkeys either fail to become sensitized or only become slightly sensitive (Bowser, Kirschstein, and Baer 1964). It should be pointed out that in large doses alkylcatechols with long side chains

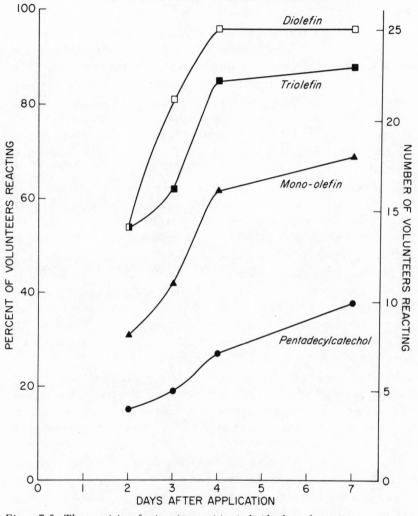

Figure 7.5. *The reactivity of poison ivy sensitive individuals to the major components of poison ivy urushiol.*
From Johnson et al. (1972).

are primary irritants and thus can cause some dermatitis at the site of contact (Baer et al. 1967).

Studies in guinea pigs have shown that young animals become more sensitive than old ones (Baer and Bowser 1963) and that the degree of sensitization achieved is greater when sensitization is attempted in winter than when it is attempted in summer (Baer and Hooton 1976). The route of administration and the vehicle are also of great importance. For example, if PDC is applied to the skin in oil or acetone solution, guinea pigs are sensitized, whereas injection subcutaneously in oil or feeding in oil not only fails to sensitize but induces immune tolerance so that the animals can no longer be sensitized (Bowser and Baer 1963).

The reactivity of purified individual components of poison ivy urushiol on the skin of naturally sensitive human volunteers has been studied (Johnson et al. 1972). The saturated component 3-n-pentadecylcatechol (PDC) (see figure 7.4, I), which from table 7.1 either is present in low concentration or is absent in poison ivy urushiol samples, reacted on less than 40 percent of the sensitive individuals (see figure 7.5). In contrast, the major poison ivy urushiol component 3-n-pentadec-8′,11′-dienyl-catechol (see figure 7.4, III), a diolefin, reacted in 25 of the 26 volunteers (see figure 7.5). The mono-olefin, 3-n-pentadec-8′-enyl-catechol (see figure 7.4, II) and the tri-olefin 3-n-pentadec-8′,11′,14′-trienyl-catechol (see figure 7.4, IV) were of intermediate toxicity (see figure 7.5).

Another interesting result from the same study is that individuals who had had their last attack of poison ivy within the past five years reacted to the test doses in 1 to 2 days after application, while those whose last attack had been 6 to 20 years prior to the test doses required 5 to 7 days to exhibit full reactivity (Johnson et al. 1972).

Since the effects of poisonous Anacardiaceae are such a scourge, means of preventing their occurrence have been eagerly sought. Although there are many preparations derived from the plants that can be swallowed or injected and are claimed to prevent the disease, there are very few well-controlled scientific studies bearing on this problem. The few controlled studies indicate that sensitivity can be reduced, but even in the most successful cases this requires massive treatment for several months or even years (Kligman 1958; Epstein et al. 1974). If severe dermatitis resulting from contact with the poisonous Anacardiaceae is to be prevented in humans, a new approach may be needed. Perhaps the induction of immune tolerance in individuals prior to their becoming exposed may prove to be useful. This approach is very successful in guinea pigs (Baer and Hooton 1976; Baer et al. 1977) and has had a limited trial in humans (Epstein 1974).

REFERENCES

Baer, H. and R. Bowser. 1963. "Antibody Production and Development of Contact Sensitivity in Guinea Pigs of Various Ages." *Science* 140:1211–12.

Baer, H., R. Watkins, A. Kurtz, J. Bycks, and C. Dawson. 1967. "Direct Contact Sensitivity to Catechols. 2. Cutaneous Toxicity of Catechols Related to the Active Principles of Poison Ivy." *J. Immunol.* 99: 365–69.

Baer, H. and M. Hooton. 1976. "Effect of Season of Immunization on the Induction of Delayed Contact Sensitivity in the Guinea Pig." *Int. Arch. Allergy Appl. Immunol.* 51:140–43.

Baer, H., M. Hooton, C. Dawson, and D. Lerner. 1977. "The Induction of Immune Tolerance in Delayed Contact Sensitivity by the Use of Chemically Related Substances of Low Immunogenicity." *J. Invest. Dermatol.* 69:215–19.

Becker, L. and G. Skipworth. 1975. "Gingko-Tree Dermatitis." *J. Am. Med. Assoc.* 231:1162–63.

Bowser, R. and H. Baer. 1963. "Contact Sensitivity and Immunologic Unresponsiveness in Adult Guinea Pigs to a Component of Poison Ivy Extract, 3-N-Pentadecylcatechol." *J. Immunol.* 91:791–94.

Bowser, R., R. Kirschstein, and H. Baer. 1964. "Contact Sensitivity in Rhesus Monkeys to Poison Ivy Extracts and Fluorodinitrobenzene." *Proc. Soc. Exp. Biol. Med.* 117:763–66.

California Department of Health. 1973. *Occupational Disease in California, 1973.* Berkeley: California Department of Health, Occupational Health Section and Center for Health Statistics.

Corbett, M. and S. Billets. 1975. "Characterization of Poison Oak Urushiol." *J. Pharm. Sci.* 64:1715–18.

Dawson, C. 1956. "The Chemistry of Poison Ivy." *Trans. N.Y. Acad. Sci. II.* 18:427–43.

Epstein, W. L. 1959. "Rhus Dermatitis." *Pediatr. Clin. North Am.* 6:843–52.

—— 1974. "Poison Oak and Poison Ivy Dermatitis as an Occupational Problem." *Cutis* 13:544–48.

Epstein, W., H. Baer, C. Dawson, and R. Khurana. 1974. "Poison Oak Hyposensitization." *Arch. Dermatol.* 109:356–60.

Gillis, W. 1971. "The Systematics and Ecology of Poison Ivy and Poison Oaks." *Rhodora* 73:72–159, 161–237, 370–443, 465–540.

Godfrey, H. and H. Baer. 1971. "The Effect of Excision of the Site of Application on the Induction of Delayed Contact Sensitivity." *J. Immunol.* 107:1643–46.

Godfrey, H., H. Baer, and R. Watkins. 1971. "Delayed Contact Sensitivity of Catechols. 5. Absorption and Distribution of Substances Related to Poison Ivy Extracts and Their Relation to the Induction of Sensitization and Tolerance." *J. Immunol.* 106:91–102.

Gross, M., H. Baer, and H. Fales. 1975. "Urushiols of Poisonous Anacardiaceae." *Phytochemistry* 14:2263–66.

Johnson, R., H. Baer, C. Kirkpatrick, C. Dawson, and R. Khurana. 1972. "Comparison of the Contact Allergenicity of the Four Pentadecylcatechols Derived from Poison Ivy Urushiol in Humans." *J. Allergy Clin. Immunol.* 49:27–35.

Kligman, A. 1958. "Poison Ivy (Rhus) Dermatitis." *Arch. Dermatol.* 77:149–80.

McNair, J. 1923. *Rhus Dermatitis.* Chicago: University of Chicago Press.

Stein, J., ed. 1967. *Random House Dictionary of the English Language.* New York: Random House.

8 Contact Hypersensitivity and Photodermatitis Evoked by Compositae

G. H. Neil Towers

In North America allergic contact dermatitis is usually associated with members of the Anacardiaceae, such as *Toxicodendron radicans* (poison ivy), *T. diversilobum* (western poison oak), and *T. vernix* (poison sumac). The sensitizing compounds are alkylated dihydroxy phenols, and they also occur in Japanese lacquer (*T. verniciferum*), mango (*Mangifera indica*), cashew nut shell oil (*Anacardium occidentale*), and Indian marking nut (*Semecarpus* spp.). Goldstein (1968) has presented an interesting account of cross-sensitivity to members of the Anacardiaceae.

The subject of contact allergy from these and other plants has been reviewed recently by Mitchell (1975). Allergic contact dermatitis, also known as delayed hypersensitivity or type IV cell-mediated hypersensitivity (Roitt 1971), is produced by contact of the skin with certain low molecular weight chemicals that sensitize blood lymphocytes. In a sensitized individual the appearance of dermatitis on reexposure to the chemical is delayed in onset by one to two days. The clinical manifestations are erythematous maculopapular rashes or papulovesicular eruptions on exposed areas of the body. Sensitization to a specific chemical is usually determined by a patch test.

Dermatitis from Sesquiterpene Lactones

Our collaborative studies at the University of British Columbia* originated with cases of allergic contact dermatitis evoked by species of *Frulla-*

* J. C. Mitchell, Division of Dermatology; G. F. Q. Chan, Faculty of Pharmaceutical Sciences; E. Camm, G. Dupius, E. Rodriguez, G. H. N. Towers, and C-K. Wat, Department of Botany; all at the University of British Columbia, Vancouver, Canada.

nia, a genus of epiphytic liverworts, in forest workers. Over the years cases have been reported in France and the Pacific Northwest (LeCoulant and Lopes 1956, 1960, 1966; Mitchell et al. 1969, 1971a; Storrs, Mitchell, and Rasmussen 1976; Bleumink et al. 1976). A simultaneous chemical investigation of the allergens, mainly by the Ourisson group in France, led to the identification of the major sensitizer (Knoche et al. 1969; Perold, Muller, and Ourisson 1972; Mitchell et al. 1970). This turned out to be a sesquiterpene lactone, which was named frullanolide (see figure 8.1, I). This com-

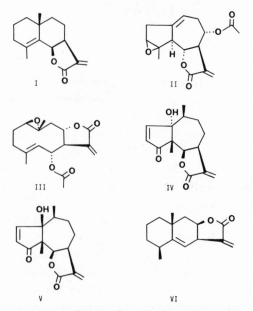

Figure 8.1. Some sesquiterpene lactones from the family Compositae. I, Frullanolide; II, Arteglasin-A; III, Pyrethrosin; IV, Parthenin; V, Hymenin; VI, Alantolactone.

pound was found to be accompanied by other, closely related sesquiterpene lactones, such as costunolide and β-arbusculin, which were also shown to be allergenic (Perold, Muller, and Ourisson 1972; Green, Muller, and Ourisson 1972; Asakawa et al. 1976). Sesquiterpene lactones are low molecular weight, colorless, bitter, lipophilic constituents of plants, and more than 500 have been identified from many species, chiefly the Compositae (Devon and Scott 1972; Yoshioka, Mabry, and Timmermann 1973). Several of them have been shown to be biologically active, with, for example, antitumor, cytotoxic, and antibiotic activities (Rodriguez, Towers, and Mitchell 1976). Their cytotoxicity is often associated with the presence in the molecule of an α-methylene-γ-lactone (Kupchan, Eakin, and Thomas 1971; Lee et al. 1971) or an α,β-unsaturated ketonic moiety (Lee et al. 1971; Lee, Meck, and Piantodosi 1973).

So far the major sources of known sesquiterpene lactones are composite species (Yoshioka, Mabry, and Timmermann 1973; Devon and Scott 1972). After preliminary results on *Frullania*, we initiated a larger study on dermatitis from Compositae. One of the more than fifty forest workers in British Columbia who had allergic contact dermatitis from *Frullania* was also discovered to be contact sensitive to *Chrysanthemum* × *morifolium* (Mitchell et al. 1970). Dermatitis caused by certain composites (e.g., chrysanthemum, ragweed, and tansy) has been known for many years (Nightingale 1931; Greenhouse and Sulzberger 1933; Mitchell et al. 1971b). Dermatitis caused by *Chrysanthemum* × *morifolium* is the commonest reported type of dermatitis from horticultural composites (Mitchell et al. 1970, 1971b). In West Germany, for instance, the plant is one of the most common causes of occupational contact dermatitis among gardeners and florists (Hausen and Schultz 1975, 1976). One of the identified allergens was shown to be arteglasin-A (see figure 8.1, II), a sesquiterpene of the guaianolide type (Hausen and Schultz 1975). About sixty species of composites, including well-known genera such as *Ambrosia, Artemisia, Aster, Cosmos, Dahlia, Helianthus, Hieracium, Matricaria, Rudbeckia, Solidago, Tagetes,* and *Xanthium,* have also been reported to cause allergic contact dermatitis (Mitchell 1970). Sesquiterpene lactones are distributed among all these groups, with the exception of *Tagetes*. In an extensive study it was shown that a prerequisite for activity is a lactone moiety with an exocyclic α-methylene function (Mitchell et al. 1970). Reduction of the methylene to a methyl group resulted in loss of activity (Mitchell et al. 1970). The allergenic properties of the sesquiterpene lactones, in other words, are correlated with their cytotoxic activities, as is shown by the work of Kupchan and associates (1971).

Cross-sensitivity patterns to patch tests with crude plant extracts vary in sensitized individuals, and in general, highly sensitized individuals are found to show a wider spectrum of cross-sensitivity than weakly sensitized persons (Mitchell et al. 1970; Hjorth and Roed-Petersen 1976). Cross-sensitivity to specific plants of the Compositae is apparently a reflection of cross-sensitivity to specific sesquiterpene lactones, but the reasons for these differences between individuals are not obvious. Very few studies have been made of this interesting and complex phenomenon.

Pyrethrum, also known as Dalmatian or Persian insect powder, consists of dried flower heads of *Chrysanthemum cinerariaefolium* and *C. coccineum* Willd. It is an effective insecticide and is used in scabiecidal and anti-mosquito preparations. It may cause allergic eczematous contact dermatitis (McCord, Kilker, and Minster 1921; Martin and Hester 1941; Sulzberger and Weinberg 1930; Sequiera, 1936; *Br. Med. J.*, 1936; Switzer 1936; Lord and Johnson 1947; McCord 1962; Mitchell, Dupuis, and Towers 1972). In a patient with allergic contact dermatitis from pyrethrum, pyrethrosin (see figure 8.1, III), a sesquiterpene lactone that does not have insecticidal properties, was shown to be the principal allergen of pyrethrum derived from species of *Chrysanthemum* (Mitchell, Dupuis, and Towers 1972).

"Australian bush dermatitis" is a chronic dermatitis of exposed areas occurring in inhabitants of the Australian bush (Burry et al. 1973). Although the dermatitis appears to be evoked by Compositae, as is evidenced by patch tests to ragweed (*Ambrosia*) and other composites, the causative species is not always known. Cases of dermatitis caused by the composites *Arctotheca*, *Cassinea*, *Inula*, and *Artemisia* are often referred to as bush dermatitis in Australia.

The Case of *Parthenium hysterophorus*

In 1974 we had reached a stage in our dermatological and phytochemical investigations of weed dermatitis at which we could confidently pinpoint the chemical source of delayed hypersensitivity to Compositae as the sesquiterpene lactones contained in these plants. Dr. Arvind Lonkar, a dermatologist from Poona, India, drew our attention to the high and increasing incidence of allergic contact dermatitis from a weed, new to that city, called *Parthenium hysterophorus*. This member of the Compositae is endemic to the Caribbean islands, Central America, the southern United States, and parts of Argentina, Brazil, and Bolivia. It is an aggressive weed of disturbed sites, and within the past 100 years it has found its way to Africa, Australia, and Asia. It has spread to a serious extent in India, where it is a major agricultural problem as well as a medical hazard (Towers et al. 1977a).

Parthenium hysterophorus was first noticed in Poona as an adventive in 1956 (Rolla 1956), and increasing numbers of cases of dermatitis were traced to exposure to it (Lonkar and Jog 1972). The dermatitis affected primarily the exposed skin surfaces (i.e., surfaces not usually covered by clothing) of agricultural workers. An increasing number of city dwellers became affected as the plant spread into urban regions (Lonkar, Mitchell, and Calnan 1974).

Contact allergy to *Parthenium* develops from repeated contacts with plants or, possibly, disseminated plant parts; after sensitization, itching eruptions occur on exposed parts of the body, particularly the upper eyelids, the sides of the neck, parts of the face, the "V" of the neck, and the fronts of elbows and the backs of knees (Lonkar, Mitchell, and Calnan 1974). The skin shows vesiculation with exudation and intense pruritus. As the dermatitis progresses, lichenification, impetiginization, fissuring, and various pigmentary changes of the skin follow.

Allergic contact dermatitis from *Parthenium* had been known in rural workers in the southern United States for many years, but with urbanization and mechanization of farming practices it has declined in incidence, according to an experienced Texas dermatologist (Howell, personal communication).

In India cases of *Parthenium* dermatitis established clinically and through patch tests have so far been reported from Poona (Lonkar, Mitchell, and Calnan 1974), Bangalore (Subba Rao et al. 1976), and Delhi (Pasricha and Shivpuri, personal communications). Suspected cases of dermatitis from *Parthenium* have also been reported anecdotally from other places, particularly in the states of Maharashtra and Karnataka, where the weed is widespread. Initially, typical patients were adult males engaged in outdoor work, particularly farmers. In recent years white-collar workers, such as bankers, doctors, police officers, and the like, have developed the allergy (Subba Rao et al. 1976). A predilection of the dermatitis for adult males has been observed in America and India and is so far unexplained. Children are spared before puberty (Lonkar, Mitchell, and Calnan 1974), and no cases of *Parthenium* dermatitis were encountered in either women or children in the course of clinical studies in Bangalore (Subba Rao, personal communication).

The major sesquiterpene lactone in most populations of *P. hysterophorus* is the pseudoguaianolide parthenin (see figure 8.1, IV). Some plant populations from southern Texas and those from southern Bolivia and central Argentina contain the diastereomer hymenin (see figure 8.1, V) as the major lactone (Rodriguez, 1975). Of ten Indian patients sensitized to *Parthenium* and to parthenin, none reacted to hymenin (Subba Rao, Rodriguez, and Towers, unpublished results), indicating that it is not only the α-methylene lactone moiety that is important but also the configuration about the bridge carbon bearing a hydroxyl group. This would suggest that the double bond in the cyclopentenone ring may participate in the allergenic reaction(s).

Helenalin and tenulin, sesquiterpene lactones that bear cyclopentenone moieties, undergo a Michael-type addition with the sulfhydryl groups of reduced glutathione and L-cysteine (see figure 8.2) (Lee et al. 1977). In the

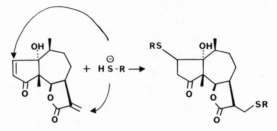

Figure 8.2. Michael-type addition of SH-compound (e.g., cysteine or a protein) to parthenin.

case of tenulin, the α-methylene-γ-lactone system also acts as an alkylating center. We have found that parthenin forms at least two adducts with cysteine under relatively mild conditions (Picman et al., unpublished results). In one of the these adducts, it is the α-methylene function that is involved; in the other, both the α-methylene and the double bond of the cyclopentenone ring. With sesquiterpene lactones that do not contain cyclopentenone rings or other reactive centers, e. g., alantolactone (see Figure 8.1, VI), it is the α-methylene group of the lactone ring that reacts with sulfhydryl groups and is presumably involved in the allergic reaction. Obviously, a great deal of study of the interaction of these compounds with membrane proteins of skin cells is needed.

Parthenin is located in the trichomes (plant hairs) of the leaves and stems (Rodriguez et al. 1976). These are easily broken off, particularly when the plants are dry and brittle, and obviously this is one cause of aerial dissemination of the sensitizer. In some ways Parthenium dermatitis resembles ragweed (Ambrosia) dermatitis in having an airborne pattern of distribution. In the case of ragweed, the allergenic determinant is in the oleoresin of the pollen (Hjorth, Roed-Petersen, and Thomsen 1976). Small amounts of parthenin are obtained in chloroform extracts of pollen of *P. hysterophorus* (Mangala and Subba Rao, unpublished results). According to Ranade (1976), the pollen is responsible for the eczematoid dermatitis. As *Parthenium* pollen is sticky, however, forming clumps in the flower heads, it may not be easily windborne, and this is one explanation for the relative unimportance of *Parthenium* pollen in hay fever or allergic rhinitis (Kahn and Grothaus 1936).

The weed has spread to nearly all the states of India, and in cities such as Poona, Hubli, and Bangalore it occupies vacant lots, ditches, roadsides, and so forth. According to a survey by the State Agriculture Department (Towers et al. 1977a), it occupies nearly one-third of the 122 sq km of the city area of Bangalore. People of India come in direct contact with the plant in various ways: working in fields, using vacant lots as toilets, using the dried plant as fuel, clearing lots or gardens, or uprooting it manually. There are no epidemics of eczematous dermatitis in Havana, Cuba, Kingston, Jamaica, Port-of-Spain, Trinidad, or many southern U.S. and Mexican towns, where the weed is prolific and conspicuous (personal observations). This must be a reflection of the habits of the people; certainly it does not seem to be genetic, because in a country such as Trinidad, where *Parthenium* is abundant and 40–50 percent of the country is of East Indian origin, there is a very low incidence of the allergy.

Parthenium hysterophorus has invaded food and fodder crop fields in India, in addition to forest nurseries. In areas that are particularly heavily infested, buffaloes and goats may graze on the weed, giving rise to the problem of "bitter milk." The weed has been shown to be quite toxic to buffaloes and cattle when fed at a 5 percent level in fodder. Autopsies reveal necrosis and lesions of the liver and gastrointestinal tract (Narasimhan et al., unpublished results). Public awareness of the problem has increased, and eradication programs have been undertaken in the more badly affected parts of the country.

It should be borne in mind that compounds other than sesquiterpene lactones may be involved in some cases of allergic contact dermatitis. For instance, pyrethrin II, an ester of chrysanthemumic acid, is a known sensitizer from *Chrysanthemum* species (Mitchell, Dupuis, and Towers 1972). Similarly, *Tagetes minuta* (Mexican marigold), a common weed in southeastern Africa, has a vesicant primary irritant effect on intact skin and can cause severe and prolonged contact dermatitis (Verhagen and Nyaga 1974). So far sesquiterpene lactones have not been found in the genus, or even in other members of the tribe Tageteae (Rodriguez 1975).

Acetylenic Compounds and Photodermatitis

Photodermatitis (dermatitis produced by a chemical, applied to skin, that is damaging on exposure to sunlight) bears a marked resemblance to *Parthenium* dermatitis (Lonkar, personal communication). It is well known that allergic contact dermatitis caused by composites is hard to distinguish from phytophotodermatitis (Curwen and Jilson 1960). For example, ragweed (*Ambrosia*) dermatitis has been misdiagnosed as a photodermatitis (Hjorth, Roed-Petersen, and Thomsen 1976). Is there any relationship between allergic contact dermatitis and photodermatitis evoked by Compositae? We believe that there may well be in certain instances and that more careful studies are needed. This leads to a discussion of photoactive chemicals in the Compositae.

There exist a number of types of compounds in plants that damage human skin in the presence of light, the best known of these being the linear furanocoumarins of the Apiaceae, Rutaceae, Papilionaceae, and Moraceae (Pathak, Kramer, and Fitzpatrick 1974). In sunlight or artificial sources of long-wave ultraviolet light (360–370 nm), these compounds cause cell damage, and the current hypothesis involves covalent cross-linking between double-stranded DNA and the furanocoumarin in photochemical reactions (Musajo et al., 1974).

Furanocoumarins or psoralens are phototoxic not only to human skin but also to bacteria and fungi (Fowlks, Griffith, and Oginsky 1958; Daniels 1965). A chance discovery by Daniels (1965), a dermatologist, that the achenes of the golden marigold (*Tagetes*) contain a chemical or chemicals phototoxic to the yeast *Candida albicans* led us to carry out a phytochemical investigation of *Tagetes* for the active substance or substances. Two of the phototoxic compounds isolated from *Tagetes patula* were identified as the thiophene derivatives α-terthienyl (see Figure 8.3, I) and 5-(3-buten-l-ynyl)-2,2'-bithienyl (see Figure 8.3, II) (Chan, Towers, and Mitchell 1975). These compounds have previously been shown to be phototoxic to the nematode *Pratylenchus penetrans* (Gommers and Geerligs 1973).

Our study was extended to 80 other species of composites (Camm, Towers, and Mitchell 1975) and more recently to 300 North, Central, and South American species (Towers et al. 1977b). Leaves, stems, roots, and achenes of each species were assayed separately using *Candida albicans* as test organism. There was excellent correlation between the reported occur-

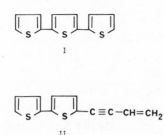

Figure 8.3. Some thiophene derivatives isolated from composites. I, α-terthienyl; II, 5-(3-buten-1-ynyl)-2,2'-bithienyl.

rences (Towers et al. 1977b) of certain polyacetylenes and their thiophene derivatives and phototoxic activity. The activity, which was usually restricted to stem, root, or achene, was sometimes even restricted to the pappus of the achene, as in *Tagetes*. Activity was correlated with the occurrence of compounds such as the tridecapentaynene $[CH_3-(C{\equiv}C)_5-CH=CH_2]$ or esters of matricarionol $[CH_3-CH=CH-(C{\equiv}C)_2-CH=CH-CH_2\,OCOR]$. These compounds, as well as a number of other polyacetylenes and thiophenes, were shown to be active. Carbonyl compounds conjugated with a double bond and further conjugated with triple bonds are phototoxic. When this conjugation is lacking, the compounds are inactive. It is possible that the mode of action of the polyacetylenes is similar to the furanocoumarins (i.e., a light-mediated cross-linking of double-stranded DNA requiring two reactive sites in a given polyacetylene molecule). However, this remains to be studied.

We have recently found that α-terthienyl can evoke photodermatitis in human skin (Chan, Towers, and Mitchell 1977). Thus, burning pain and erythema developed within 20 minutes of long-wave ultraviolet irradiation of α-terthienyl skin sites, in contrast to α-methoxypsoralen-evoked photodermatitis, which developed at about 48 hours (Chan, Towers, and Mitchell, 1977). α-Terthienyl and related thiophenes occur in many composites, particularly in the tribes Helenieae, Heliantheae, and Tageteae. In *Tagetes* the highest concentrations are in the roots, but these compounds also occur in the flower (Glushka et al., unpublished results). *Tagetes minuta*, as mentioned previously, is an adventive weed in East Africa and, apart from its other undesirable properties, causes severe irritation and edema of small wounds such as scratches (Verhagen and Nyaga 1974). Walking through a field of marigolds on a sunny afternoon could therefore be a hazard. One of us (Towers) found that exposure to the tropical sun of an area of the forearm that had been treated with a 1 percent solution of α-terthienyl in petrolatum resulted in severe erythema within a few minutes and also in hyperpigmentation that persisted for two months.

So far, polyacetylenes such as $C_6H_5-(C\equiv C)_3-CH_3$ or $CH_3-(CH\equiv CH)_5-CH=CH_2$, which we have established as being phototoxic to *Candida* and certain pathogenic bacteria, have not been tested on human skin. It is more than likely that they have an effect, and this could be important if these compounds are carried in pollen. Many composites contain polyacetylenes (Bohlmann, Burkhardt, and Zdero 1973) as well as sesquiterpene lactones, and thus may be capable of a dual "attack" on exposed skin. The possibility of photodermatitis from Compositae is worth investigating, especially in cases in which eczematoid dermatitis bears close resemblance to photodermatitis and in which the offending species is known to contain reactive sesquiterpene lactones as well as photoactive polyacetylenes.

REFERENCES

Asakawa, Y., J. Muller, G. Ourisson, J. Fousserau and G. Ducombs. 1976. "Nouvelles lactones sesquiterpéniques de Frullania (Hepaticae)." *Bull Soc. Chim. Fr.* 1976:1465–66.

Bleumink, E., J. Mitchell, T. Geissman and G. Towers. 1976. "Contact Hypersensitivity to Sesquiterpene Lactones." *Contact Dermatitis* 2:81–88.

Bohlmann, F., T. Burkhardt and C. Zdero. 1973. *Naturally Occurring Acetylenes.* London: Academic Press.

Br. Med. J. 1936. "Pyrethrum Dermatitis" (editorial). 1936:1262.

Burry, J., R. Kuchel, J. Reid and J. Kirk. 1973. "Australian Bush Dermatitis: Compositae Dermatitis in South Australia." *Med. J. Aus.* 1:110–16.

Camm, E., G. Towers and J. Mitchell. 1975. "UV-Mediated Antibiotic Activity of Some Compositae Species." *Phytochemistry* 14:2007–11.

Chan, G., G. Towers and J. Mitchell. 1975. "Ultraviolet-Mediated Antibiotic Activity of Thiopene Compounds of *Tagetes*." *Phytochemistry* 14:2995–96.

—— 1977. "Photodermatitis Evoked by α-Terthienyl." *Contact Dermatitis* 3:215–16.

Curwen, W. and O. Jilson. 1960. "Light Hypersensitivity." *J. Invest. Dermatol.* 34:207–12.

Daniels, F. 1965. "A Simple Microbiological Method for Demonstrating Phototoxic Compounds." *J. Invest. Dermatol.* 44:259–63.

Devon, T. and A. Scott. 1972. *Handbook of Naturally Occurring Compounds*, vol. 2. *Terpenes.* London: Academic Press.

Fowlks, W., D. Griffith and W. Oginsky. 1958. "Photosensitization of Bacteria by Furocoumarins and Related Compounds." *Nature (London)* 181:571–72.

Glushka, J., G. Towers, G. Chan and J. Mitchell. Unpublished results.

Goldstein, N. 1968. "The Ubiquitous Urushiols—Contact Dermatitis from Mango, Poison Ivy and Other 'Poison' Plants." *Cutis* 4:679–85.

Gommers, F. and J. Geerligs. 1973. "Lethal Effect of Near Ultraviolet Light on *Pratylenchus penetrans* from Roots of *Tagetes*." *Nematologia* 19:389–93.

Green, A., J. Muller and G. Ourisson. 1972. "A New Approach to α-Methylene-γ-Butyrolactones. Synthesis of (–)-Frullanolide." *Tetrahedron Lett.* 1972:2489–92.

Greenhouse, C. and M. Sulzberger. 1933. "Common Weed (*Tanacetum vulgare*) as a Cause of Eczematous Dermatitis." *J. Allergy* 4:523–26.

Hausen, B. and K. Schulz. 1975. "A First Allergenic Sesquiterpene Lactone from *Chrysanthemum indicum* L., Arteglasin-A." *Naturwissenschaften* 62:585–86.

—— 1976. "Chrysanthemum Allergy 3. Identification of the Allergens." *Arch. Dermatol. Res.* 255:11–21.

Hjorth, N. and J. Roed-Petersen. 1976. "Compositae Sensitivity Among Patients with Contact Dermatitis." *Contact Dermatitis* 2:271–81.

Hjorth, N., J. Roed-Petersen and K. Thomsen. 1976. "Airborne Contact Dermatitis from Compositae Oleoresins Simulating Photodermatitis." *Br. J. Dermatol.* 95:613–19.

Kahn, I. and E. Grothaus. 1936. "Parthenium Hysterophorus: Antigenic Properties, Respiratory and Cutaneous." *Tex. State J. Med.* 32:284–87.

Knoche, H., G. Ourisson, G. Perold, J. Fousserau and J. Maleville. 1969. "Allergenic Component of a Liverwort: A Sesquiterpene Lactone." *Science* 166:139–40.

Kupchan, S., M. Eakin and A. Thomas. 1971. "Tumor Inhibitors. 69. Structure–Cytotoxicity Relationships Among the Sesquiterpene Lactones." *J. Med. Chem.* 14:1147–52.

LeCoulant, M. and G. Lopes. 1956. "A propos des dermites des bois." *J. Med. Bordeaux.* 3:245–49.

—— 1960. "Role pathogène des muscinées hepatiques dans les industries du bois." *Société de Medicine et d'Hygiène du Travail* 21:374–76.

—— 1966. "L'allergie au *Frullania*; son role dans la 'Dermite du bois de chêne.'" *Bull. Soc. Fr. Dermatol. Syphiligr.* 73:440–43.

Lee, K., E. Huang, C. Piantadosi, J. Pagano and T. Geissman. 1971. "Cytotoxicity of Sesquiterpene Lactones." *Cancer Res.* 31:649–54.

Lee, K., R. Meck and C. Piantadosi. 1973. "Antitumor Agents. 4. Cytotoxicity and *in Vivo* Activity of Helenalin Esters and Related Derivatives." *J. Med. Chem.* 16:299–301.

Lee, K., I. Hall, E-C. Mar, C. Starnes, S. El Gebaly, T. Waddell, R. Hadgraft, C. Ruffner and I. Weidner. 1977. "Sesquiterpene Antitumor Agents: Inhibitors of Cellular Metabolism." *Science* 196:533–36.

Lonkar, A. and M. Jog. 1972. " 'Epidemic' Contact Dermatitis from *Parthenium hysterophorus*." *Contact Dermatitis Newsletter, London* 11:291.

Lonkar, A., J. Mitchell and C. Calnan. 1974. "Contact Dermatitis from *Parthenium hysterophorus*." *Trans. St. John's Hosp. Dermatol. Soc.* 60:43–53.

Lord, K. and C. Johnson. 1947. "Production of Dermatitis by Pyrethrum and Attempts to Produce Non-irritant Extract." *Br. J. Dermatol.* 59:367–75.

Mangala, A. and P. Subba Rao. Unpublished reults.

Martin, J. and K. Hester. 1941. "Dermatitis Caused by Insecticidal Pyrethrum Flowers." *Br. J. Dermatol.* 53:127–42.

McCord, C., C. Kilker and D. Minster. 1921. "A Record of Occurrence of Occupational Dermatoses Among Workers in the Pyrethrum Industry." *J. Am. Med. Assoc.* 77:448–49.

McCord, C. 1962. "The Occupational Toxicity of Cultivated Flowers." *Ind. Med. Surg.* 31:365–68.

Mitchell, J., W. Schofield, B. Singh and G. Towers. 1969. "Allergy to *Frullania*: Allergic Contact Dermatitis Occurring in Forest Workers Caused by Exposure to *Frullania nisquallensis.*" *Arch. Dermatol.* 100:46–49.

Mitchell, J. 1970. "Allergic Contact Dermatitis from Compositae." *Trans. St. John's Hosp. Derm. Soc.* 55:174–83.

Mitchell, J., B. Fritig, B. Singh and G. Towers. 1970. "Allergic Contact Dermatitis from *Frullania* and Compositae. The Role of Sesquiterpene Lactones." *J. Invest. Dermatol.* 54:233–39.

Mitchell, J., A. Royal, G. Dupuis and G. Towers. 1971a. "Allergic Contact Dermatitis from Ragweeds (*Ambrosia* Species)." *Arch Dermatol.* 104:73–76.

Mitchell, J., T. Geissman, G. Dupuis and G. Towers. 1971b. "Allergic Contact Dermatitis Caused by *Artemisia* and *Chrysanthemum* Species." *J. Invest. Dermatol.* 56:98–101.

Mitchell, J., G. Dupuis and G. Towers. 1972. "Allergic Contact Dermatitis from Pyrethrum (*Chrysanthemum* spp.). The Roles of Pyrethrosin, a Sesquiterpene Lactone, and of Pyrethrin II." *Br. J. Dermatol.* 86:566–73.

Mitchell, J. 1975. "Contact Allergy from Plants." In V. Runeckles, ed., *Recent Advances in Phytochemistry*, 9:119–38. New York: Plenum Press.

Musajo, L., G. Rodighiero, G. Caporale, F. Dall'Acqua, S. Marciani, F. Bordin, F. Bacichetti and R. Bevilacqua. 1974. "Photoreactions Between Skin-Photosensitizing Furocoumarins and Nucleic Acids." In T. Fitzpatrick, ed., *Sunlight and Man*, pp. 369–88. Tokyo: University of Tokyo Press.

Narasimhan, T., M. Ananth, M. Babu, A. Mangala and P. Subba Rao. Unpublished results.

Nightingale, G. 1931. "*Chrysanthemum* Dermatitis." *Lancet* 1:1132–34.

Pathak, M., D. Kramer and T. Fitzpatrick. 1974. "Photobiology and Photochemistry of Furocoumarins (Psoralens)." In T Fitzpatrick, ed., *Sunlight and Man*, pp. 335–68. Tokyo: University of Tokyo Press.

Perold, G., J. Muller and G. Ourisson. 1972. "Structure d'une lactone allergisante: le frullanolide-1." *Tetrahedron* 28:5797–5803.

Picman, A., G. Towers, E. Rodriguez, P. Jamieson and E. Piers. Unpublished results.

Ranade, S. 1976. "Parthenium—A Positive Danger." In *Seminar Proceedings*, p. 15. Bangalore, India: BICRO and University of Agricultural Sciences, Bangalore.

Rodriguez, E. 1975. "The Chemistry and Distribution of Sesquiterpene Lactones and Flavonoids in *Parthenium* (Compositae): Systematic and Ecological Implications." Ph.D. dissertation, University of Texas.

Rodriguez, E., G. Towers, and J. Mitchell. 1976. "Biological Activities of Sesquiterpene Lactones—A Review." *Phytochemistry* 15:1573–80.

Rodriguez, E., M. Dillon, T. Mabry, J. Mitchell and G. Towers. 1976. "Dermatologically Active Sesquiterpene Lactones in Trichomes of *Parthenium hysterophorus* (Compositae)." *Experientia* 32:236–38.

Roitt, I. 1971. *Essential Immunology*. Oxford, England: Blackwell Scientific Publications.

Rolla, S. 1956. "*Parthenium hysterophorus* Linn: A New Record for India." *J. Bombay Nat. Hist. Soc.* 54:219.

Sequiera, J. 1936. "Pyrethrum Dermatitis." *Br. J. Dermatol.* 48:473–76.

Storrs, F., J. Mitchell and M. Rasmussen. 1976. "Contact Hypersensitivity to Liverwort and the Compositae Family of Plants." *Cutis* 18:681–86.

Subba Rao, P., A. Mangala, B. Subba Rao and K. Prakash. 1976. "Parthenium—A Positive Danger." *In Seminar Proceedings*, p. 17. Bangalore, India: BICRO and University of Agricultural Sciences, Bangalore.

Subba Rao, P., E. Rodriguez and G. Towers. Unpublished results.

Sulzberger, M. and C. Weinberg. 1930. "Dermatitis Due to Insect Powder." *J. Am. Med. Assoc.* 95:111–12.

Sweitzer, S. 1936. "Scabies: Further Observations on Its Treatment with Pyrethrum Ointment." *J. Lancet* 56:467–68.

Towers, G., J. Mitchell, E. Rodriguez, P. Subba Rao and F. Bennett. 1977a. "Biology and Chemistry of *Parthenium hysterophorus* L., a Problem Weed in India." *J. Sci. Ind. Res.* 36:672–84.

Towers, G., C-K. Wat, E. Graham, R. Bandoni, C. Chan, J. Mitchell, and J. Lam. 1977b. "Ultraviolet-Mediated Antibiotic Activity of Species of Compositae Caused by Polyacetylenic Compounds." *Lloydia* 40:487–98.

Verhagen, A. and J. Nyaga. 1974. "Contact Dermatitis from *Tagetes minuta*." *Arch Dermatol.* 110:441–44.

Yoshioka, H., T. Mabry and B. Timmermann. 1973. *Sesquiterpene Lactones*. Tokyo: University of Tokyo Press.

Index

Abortifacient agents, 109, 111, 115

Abrin, 92–97; stability, 93, 96; structure, 92, 93

Abrus lectin, 90, 91, 92–97, 98, 99, 100; purification, 92; structure, 92–97

Abrus precatorius, 84, 89, 94, 99

Acetaldehyde, 36, 41

Acetaldehyde-N-methyl-N-formylhydrazone, *see* Gyromitrin

N-Acetyl-4-hydroxymethylphenylhydrazine, 40

Acids: ibotenic, 8, 25–27, 33; α-lipoic, 18; organic, 121, 128; oxalic, 110, 123; tannic, 131; thioctic, 18; tropic, 63

Africa, 29, 64, 125, 145, 174, 179

African violet, see *Saintpaulia ionantha*

Agaricus bisporus, 38

Agaritine, 38, 40

Agricultural workers, 175

Alcoholism, 41, 44

Aleurites fordii, 140, 148

Alkaloids: ingenol derivatives, 142; pyridine/piperidine, 59, 65–69; quinazolone, 118; solanum, 69–72, 74–79, 122, 131; steroidal, 59, 69–79, 122, 131; tropane, 24, 27, 34, 59, 63–65, 131, 132; unspecified type, 104, 110, 112, 114, 115, 118, 120, 121, 122, 126, 128; veratrum, 69–75, 76, 77

Alkylcatechols, 164, 165, 166, 168

Allergic rhinitis, 177

Aluminum plant, see *Pilea cadierei*

Amanin, 11, 12

Amanita: A. *bisporigera*, 9; A. *cothurnata*, 8, 27; A. *muscaria*, 8, 21–27, 32; A. *ocreata*, 9; A. *pantherina*, 8, 27; A. *phalloides*, 9, 10, 13, 18, 19, 21; A. *suballiaceaa*, 9; A. *tenuifolia*, 9; A. *verna*, 9, 13; A. *virosa*, 10, 13

Amanitins, 11, 12, 14, 16, 17, 18, 21

Amanullin, 11, 12

Amatoxin–RNA polymerase II complex, 15

Amatoxins, 7, 8, 9–19, 22, 32; dethio derivatives, 11; seco derivatives, 11; secodethio derivatives, 12; structural studies, 11–12

Amino acids, 11, 13, 18, 19, 21, 26, 33, 43, 87, 88, 92, 93, 96, 99, 176; alanine, 11, 21; γ-aminobutyric acid, 26; asparagine, 11, 19, 87; aspartic acid, 11; 19; cysteine, 176; γ,δ-dihydroxyisoleucine, 11; glutamic acid, 33; glutamine, 87; glutathione, 176; glycine, 11, 87; γ-hydroxyisoleucine, 11, 13; hydroxyproline, 11; isoleucine, 11; methionine, 18; phenylalanine, 18, 21; proline, 21; tryptophan, 11; tyrosine, 18; valine, 21

Amygdalin, 131

Anacardiaceae, 161–70, 171; botany, 161–63

Anacardium occidentale, 161, 165, 171

Analgesic agents, 64, 114, 116, 117, 121, 122, 124, 132

Anaphylaxis, 119

Antacids, 132

Antamanide, 21

Anthelmintic agents, 116, 117, 121, 126

Anthocyanin, 115

Anthurium, 133; A. *andraeanum*, 105, 106, 108, 110

Antibiotics, 173; actinomycin D, 14; chloramphenicol, 17; neomycin, 17, 126, 132; penicillin G, 17

Antibody reagents, mitogens, 84

Antidotes to plant poisoning, 18, 20, 24, 34, ‧38, 131

Antidysenteric agents, 117

Antifertility agents, 109, 125

Antihemorrhagic agents, 114, 124, 148

Antihistamine, 132; chlorpheniramine, 123; diphenhydramine, 123; promezathine, 124

Antileukemic agents, 140, 142, 148

Antirheumatic agents, 114, 119, 124

Antitumor agents, 128, 172
Antitussive agents, 109, 111
Apiaceae, 178
Araceae, *see names of genera*
Argentina, 174, 175
Arisaema, 132; *A. triphyllum*, 105, 106, 108, 110
Armillaria mellea, 10
Aroids, *see* Araceae, genera
Arrow poisons, 145, 148
Artificial respiration, 131, 132
Asia, 29, 161, 174
Aspirin, poisoning by, 3
Asthma treatment, 64
Astragalus, 61
Atropa, 59; *A. belladonna*, 63, 119, 127
Atropine, *see* Tropane alkaloids, atropine
Auditory disturbances due to plant ingestion, 22, 25
Australia, 29, 61, 174
Aztec, 28

Babylonians, 64
Bacteria, 178
Bacterial products, mitogens, 84
Bacterial toxins, 93
Baeocystin, 30
Baliospermum montanum, 140, 141, 143
Basophils, 86
Begonia, 108, 133; hybrids, 104, 105, 106, 110; *B. barbata*, 110; *B. cucullata*, 110; *B. gracilis*, 110; *B. hirtella*, 110; *B. malabarica*, 110; *B. picta*, 110; *B. rex*, 110; *B. sanguinea*, 110; *B. sutherlandii*, 110; *B. tuberosa*, 110
Benzo[α]pyrene, 151
Bile-duct cannulation, 15
Bioassay methods: acute toxicity (LD_{50}) determinations, 6, 13, 19, 26, 28, 33, 37, 112, 123, 153; Berenblum experiments, diterpene esters, 152; hallucination induction, psilocin and psilocybin, 29; hamster teratogenic assay, steroidal alkaloids, 74, 75, 77, 78, 79; insecticidal assay, *Amanita muscaria*, 25; irritant dose 50%, diterpene esters, 152; lachrymation production, *Coprinus atramentarius*, 42; narcosis-potentiating assay, *Amanita muscaria*, 25; P-388 (PS) leukemia cell culture system,

diterpene esters, 143; phototoxicity for *Candida albicans*, of composite species, 178–79; production of hyperaldehydemia, *Coprinus atramentarius*, 42; toxicity for rodents, of household plants, 104–7
Birth defects, 59, 61, 65, 67–79
"Bitter milk," 177
Black locust (*Robinia pseudoacacia*), 89, 98–100
Blighia sapida, 43
Blood: chemistry values, 15, 17; erythrocytes, 83; granulocytes, 83; leucocytes, 83; toxic effects on, by plants, 10, 21, 88, 89, 92, 98, 100, 121, 128; whole transfusions, 19
Boletus edulis, 13
Bolivia, 174, 175
Boston fern, see *Nephrolepis exaltata*
Buffaloes, 177
Burn treatment, 120, 127

Ca^{++}-ionophore, 84
Caladium, 104, 108, 109, 111, 132; *C. bicolor*, 106, 110–11; *C. colocasia*, 111; *C. picturatum*, 111
Calcinosis due to plant ingestion, 62–63
Calcium oxalate, 109, 110, 111, 112, 113, 119, 120, 123, 127, 132
Canada, 31, 173
Canavalia ensiformis, 84, 86
Candelabrum cactus, see *Euphorbia lactea*
Candida albicans, 178
Cantharellus cibarius, 13
Caoutchouc, 114, 115, 116
Capers, 177
Caper spurge, see *Euphorbia lathyris*
Carcinogenesis, chemical: human esophageal, 140; initiation, 151; mouse leukemia, 151; mouse liver, 151; mouse lung, 151; mouse skin, 137, 150–51, 152; promotion, 137, 151, 154; rat mammary, 151; tissue culture, 151; "two-stage" (Berenblum) experiments, 137, 150–51
Carcinogens, 39, 40, 137, 151, 154
Cardiac glycosides (cardiotonics), 61, 109, 112, 114, 119, 124, 127, 131; convallamarin, 109, 112; convallarin, 109, 112; convallatoxin, 112; convalloside, 112; digitoxin, 114, 124; nerioside, 119, 127; oleandroside, 119, 127

Cardiovascular system, toxic effects on due to plant ingestion, 10, 21, 32, 40, 61, 63, 72, 109, 118, 123, 124, 131
Caribbean Islands, 174
Cash crops in U.S. agriculture, 60
Cashew nut oil tree, see *Anacardium occidentale*
Cassava (*Manihot*), 145
Castor bean lectin, 96
Castor oil, 145
Cathartics, 121, 128
Cats, 120, 127
Cattle, 62, 64
Central America, 28, 109, 174
Central nervous system, toxic effects due to plant ingestion, 22, 25, 26, 28, 35, 63, 66, 72, 114, 131
Ceriman, see *Monstera deliciosa*
Cestrum, 59, *C. diurnum*, 61–62; *C. parqui*, 61
Charcoal, activated, 16, 130, 131, 132
Chemical assay methods: amatoxins, 12–13; gyromitrin and analogs, 36; muscarine, 34; oxalic acid, 123
Chickens, 64
Chloral hydrate, 64
Chlorpromazine, 31
Cholecalciferol, 62
Chromatography: adsorption, 42; affinity, 87, 88, 90, 92, 98; gas-liquid, 36, 42; ion-exchange, 42; Sephadex, 13, 42; thin-layer (TLC), 13
Chrysanthemum: *C. cinerariaefolium*, 174; *C. coccineum*, 174; *C.* × *morifolium*, 173
Cigarettes, 65
Clinical management of plant poisoning, see Treatment of plant poisoning
Clitocybe: *C. dealbata*, 34; *C. rivulosa*, 33
Cocarcinogens, see Tumor-promoters
Codiaeum, 108, 109, 111; *C. variegatum*, 111–12; *C. variegatum* var. *pictum*, 104, 106, 112
Colchicine, 86
Cold treatments, 117, 119
Colombia, 140
Commerical mushroom (*Agaricus bisporus*), 38
Compositae, 61, 171–83; tribes, 177, 179
Concanavalin A, 84, 86, 96

Conium maculatum, 68
Conocybe: *C. cyanopus*, 32; *C. filaris*, 32; *C. smithii*, 32
Contact allergens, 125, 163–69, 171–77
Contact allergy, see Dermatitis, allergic contact
Convallaria, 104, 108; *C. majalis*, 106, 109, 112, 131
Coprine, 8, 40–44
Coprinus: *C. atramentarius*, 8, 40–44; *C. comatus*, 40; *C. insignis*, 44; *C. micaceus*, 40; *C. quadrifidus*, 44; *C. variegatus*, 44
Coprinus–ethanol reaction, 40–41, 42
Coral plant (*Jatropha multifida*), 148
Corn, 60
Corticosteroids, 132; cortisone acetate, 123; dexamethasone, 126, 132
Crab's eye, see *Abrus precatorius*
Croton: *C. flavens*, 140; *C. sparsiflorus*, 139; *C. tiglium*, 137, 138, 139, 141, 148
Croton oil, 137, 138, 150, 151, 152
Crotons, see *Codiaeum*
Crown-of-thorns (*Euphorbia millii*), 142, 145, 148
Cryptic cocarcinogens, 139
Curaçao, 140
Cyanide antidote kit, 131
Cytochrome C treatment, 19
Cytotoxic agents, 121, 128, 140, 143, 173

Dalmatian insect powder, 174
Daphnane, 143; esters, 143–44, 149, 152
Daphne mezereum, 3
Datura, 59, 64; *D. stramonium*, 61–63
Day lily, see *Hemerocallis lilio-asphodelus*
Deadly nightshade (*Atropa belladonna*), 63, 119, 127
Delphinium, 61
Depilatory agent, 115
Dermatitis: allergic contact, 112, 119, 125, 161–70, 171–77; "Australian bush," 174; by composite species, 171–77; contact, 110, 111, 113, 116, 122, 128; cross-sensitivity, 170, 173; occupational, 161, 173; by *Parthenium hysterophorus*, 174–77, 178; photo-, 116, 177–80; by poisonous Anacardiaceae, 161–71; *Rhus*, 162
Detoxification, *Gyromitra esculenta*, 39–40

Index

Diarrhea treatment, 117
Diazepam, 27, 131, 132
Dieffenbachia, 104, 108, 109, 124, 132; *D. amoena*, 106, 109, 112–13; *D. bausei*, 106, 113; *D. exotica*, 123; *D. picta*, 106, 109, 13, 123; *D. seguine*, 109, 113
Digestion aids, 114, 118, 119
Digitalis, 104, 108; *D. purpurea*, 106, 113–14, 124, 131
Digitalis (drug), 114, 124
1, 25-Dihydroxyvitamin D₃ glycoside, 62
1,1-Dimethylhydrazine, 38
Dioscorides, 3
Diosgenin, 75, 76
Diphenylhydantoin, 132
Disulfiram, 8, 41, 42, 43, 44
Diterpene esters, *see* Daphnane esters; Ingenane derivatives; Phorbol derivatives
Diuretics, 108, 110
DNA, 14, 15, 86, 90, 178, 179
Domestic animals and livestock, 1, 5, 6, 59–79, 118–19, 127, 137, 153, 177
Dracaena, 108; *D. cinnabari*, 114; *D. cylindrica*, 114, 124; *D. fragrans*, 114; *D. laxissima*, 114, 124; *D. sanderiana*, 104, 106, 114, 124; *D. steudtneri*, 114
Drugs, 59
Duke, J. A., 2
Dysmenorrhea treatment, 117

East Indies, 109
Edible plants, 5, 13, 35, 39, 59, 60, 108, 109, 110, 111, 117, 126, 127, 128, 137, 148
Eggplant (*Solanum melongena*), 70, 75
Elaeophorbia: E. drupifera, 142; *E. grandiflora*, 142
Electrolyte disturbances due to plant ingestion, 10, 17, 21, 131
Emesis induction, 4, 130, 131, 132
Emetics, 72, 108, 110, 112, 116, 130
Emetine, 130
Enzymes: alkaline phosphatase, 17, 30; ceruloplasmin, 31; chymotrypsin, 123; cytochrome oxidase, 31, 131; cytochrome P-450, 39; dopamine-β-hydroxylase, 44; "dumcain," 109, 132; ficin, 116, 117; LDH, 15, 17; papain, 84, 132; proteolytic (general), 100, 109; RNA-polymerases, 12,

14, 15; SGOT, 15, 17; SGPT, 15, 17; trypsin, 84, 90, 98, 100, 123
Ethanol, 22, 40, 42
Euphorbia, 108, 132, 137, 140–54 *passim*; *E. balsamifera*, 145, 148; *E. biglandulosa*, 142; *E. coerulescens*, 139; *E. cooperi*, 139; *E. esula*, 142; *E. franckiana*, 139; *E. fortissima*, 139; *E. helioscopia*, 139; *E. ingens*, 141, 142; *E. jolkini*, 142; *E. kansui*, 141; *E. lactea*, 114, 125–26, 142, 148; *E. lathyris*, 141, 142, 148, 154; *E. millii*, 142, 145, 148; *E. myrsinites*, 142; *E. poisonii*, 139, 140, 143, 144, 145; *E. polyacantha*, 139; *E. pulcherrima*, 105, 106, 115, 124–25, 150; *E. resinifera*, 139, 140, 142, 143, 144, 148; *E. segueiriana*, 142; *E. serrata*, 142; *E. tirucalli*, 105, 106, 115–16, 125–26, 138, 139, 140, 142, 148, 152; *E. triangularis*, 139; *E. unispina*, 139, 143, 144
Euphorbia, sections of genus, 149–51
Euphorbiaceae, 137–59; botany, 144–45; subfamilies, 145; tribes, 146–47, 150
Euphorbon, 114, 116
Europe, 9, 18, 27, 29, 35, 36, 39
Excoecaria agallocha, 143
Excretion, *see* Metabolism studies
Experimental animals: brine shrimp larvae, 153; buffaloes, 177; cattle, 6, 66, 68, 69, 74, 177; dogs, 3, 15, 16, 18, 125; goats, 74, 166; hamsters, 74, 75, 76, 77, 78, 79; houseflies, 25, killifish, 153; leeches, 110; mice, 13, 14, 17, 18, 19, 28, 30, 33, 37, 38, 41, 42, 67, 75, 104, 105, 106, 107, 109, 125, 126, 127, 128, 137, 142, 150, 151, 152; monkeys, 24, 166; pigs, 65, 67–68; *Pratylenchus penetrans*, 178; rabbits, 37, 39, 74, 109, 123, 125; rats, 26, 31, 37, 42, 43, 74, 75, 104, 105, 106, 107, 123, 124, 125, 126, 127, 128, 150, 151; sheep, 66, 68, 72, 74, 166
Eye infection treatment, 120, 127

Farmers, and contact dermatitis, 175
Fat cells, 83
Fava bean lectin (favin), 84, 96
Favism, 129
Ficus, 108, 133; *F. barclayana*, 116; *F. carica*, 116; *F. cotinifolia*, 116; *F. elastica*

(*decora*), 105, 106, 116–17, 126; *F. glabrata*, 117; *F. hauili*, 117; *F. laurifolia*, 117; *F. lyrata* (*pandurata*), 104, 106, 117, 126; *F. platyphylla*, 117; *F. retusa*, 117
Fish poisons, 116, 137, 148, 153
Flavonols, 121
Flavoring agents, 111
Florists, 173
Fly agaric, see *Amanita muscaria*
Folkloric uses of poisonous plants, *see under individual biological activity*
Food additives, 5
Food plants, *see* Edible plants
Forage plants, 60, 61, 69–70, 177
Forests, 60
Forest workers, 172
Foxglove, see *Digitalis purpurea*
France, 172
Frullania, 171, 173
Fungi, 7–58, 83, 178
Furanocoumarins, 178, 179

Galactopoietic agent, 115
Gardeners, 173
Garden pea lectin, 96
Garden plants, *see* Ornamental plants
Gastric lavage, 4, 118, 129, 132
Gastrointestinal toxic effects due to plant ingestion, 8, 10, 15–16, 22, 24, 32, 35, 40, 61, 66, 72, 111, 114, 115, 116, 117, 118, 124, 126, 131, 177
Gel electrophoresis, 92, 99
Gingko tree, 165
Gladiolus, 108, 117, 126; *G. ecklonii*, 117; *G. edulis*, 117; *G. gandavensia*, 104, 106, 117, 126; *G. ludwigii*, 117; *G. multiflorus*, 117; *G. psittacinus*, 117; *G. saundersii*, 117
β-N-[γ-L(+)-Glutamyl]-4-hydroxymethyl-phenylhydrazine, *see* Agaritine
Glycoproteins, 21, 83, 87–88, 92
Glycosides, alkaloidal, 70, 71, 72, 75; cardiac, 61, 109, 112, 114, 119, 124, 127, 131; cyanogenic, 104, 111, 126, 131; unspecified, 104, 118
Goats, 177
Governmental agencies, 4, 5
Grazing lands, 60

"Grune Knollenblatterpilz," see *Amanita phalloides*
Gum euphorbium, 114
Gymnopilus spectabilis, 32
Gyromitra esculenta, 8, 35–40
Gyromitrin, 8, 35–40

Hallucinogenic plants, 28, 29, 64
Halogeton, 61
Havana, Cuba, 177
Hay fever, 177
Heart, toxic effects on due to plant ingestion, 35, 61, 131
Heidelberg University, 138
Hemerocallis, 108; *H. lilio-asphodelus* (*flava*), 105, 106, 117, 126
Hemodialysis, 18
Hemoperfusion, 17
Hemorrhoid treatment, 121
Henbane (*Hyoscyamus niger*), 63–64
Hepatic failure due to plant ingestion, 10, 15, 19, 20, 35, 38, 118, 177
Heptadecylcatechols, 164, 165
Hershey Medical Center, Pennsylvania State University, 89
High Times, 31
Hippeastrine, 118
Hippeastrum, 105, 108, 118, 126; hybrids of, 104, 105, 106, 117, 126; *H. equestre*, 118; *H. vittata*, 118
Hippomane mancinella, 140–41, 143
Histamine, 86, 123, 124
Hodgkin's disease, 85
Hogs, 64
Hoover, S., 89
Horses, 62, 64
Horticultural associations, 4
Household plants, *see* Ornamental plants
Humans, effects of plant poisoning, 3, 7, 10, 13, 16, 21, 22, 25, 26, 27, 28, 30, 32, 35, 37, 38, 39, 40, 43, 63, 64, 66, 67, 103, 109, 111, 112, 113, 114, 115, 116, 118, 119, 122, 123, 124, 125, 126, 131, 132, 153, 154, 161, 167, 168, 171, 173, 174, 175, 176, 178, 179, 180
Hura crepitans, 84, 89, 90, 98, 100, 143
Hura lectin, 90–92, 98–100; purification, 98; structural studies, 99–100
Hydrangin, 118

Hydrangea, 104, 108, 118, 126; *H. ar-borescens,* 105, 106, 118; *H. macro-phylla,* 106, 118
Hyoscyamus, 59; *H. niger,* 63–64
Hyperbaric oxygenation, 19
Hypersensitivity, delayed (= Type IV cell-mediated), *see* Dermatitis, allergic contact
Hypervitaminosis D, 62
Hypholoma fasciculare, 8
Hypoglycemia due to plant ingestion, 10, 18, 43
Hypoglycin A, 43
Hypotensive agents, 69, 72

Immune response and mitogenic stimulation, 84–85
Immune tolerance induction, 168
Immunologic disease, 161
Impotency remedy, 116
India, 22, 174, 175, 177
Indian marking nut (*Semecarpus*), 171
Induced vomition, *see* Emesis induction
Inflammation: oral, caused by plants, 108, 109, 110, 111, 112, 113, 116, 119, 123, 125; skin, caused by plants, *see* Dermatitis, contact
Information on poisonous plants, 1–6 *passim*
Ingenane, 141; derivatives, 141, 142, 149, 152
Inky cap, see *Coprinus atramentarius*
Inocybe napipes, 34
Insecticidal agents, 5, 69, 103, 116, 145, 174
Insects, plant toxins as defense against, 2
Interferon, 83
Intravenous fluids, 131, 132
Ipecac, 111, 130
Iproniazid, 39
Irritants, *see* Skin, irritants
Isoniazid, 39
Isoxazole derivatives, *see* Acids, ibotenic; Muscimol

Jack-in-the-pulpit, see *Arisaemia triphyllum*
Japan, 24, 163
Japanese lacquer tree (*Toxicodendron ver-niciferum*), 163, 171
Jatropha multifida, 148

Jerusalem cherry, see *Solanum pseud-ocapsicum*
Jimsonweed (*Datura stramonium*), 61–63

Kaufman, S., 89
Kingston, Jamaica, 177
Kinins, 109, 112

Labeling of poisonous ornamental plants, 4–5
Lantana, 104, 108; *L. camara,* 106, 118–19, 126
Lectin–carbohydrate receptor complexes, 85
Lectins, 83–102; aggregation, 85; properties, 83; purification, 85, 89
Lens culinaris (lentil), 84
Lignans, 121
Liliaceae, 59–82, 104, 105, 106, 108, 117, 124, 126
Lily-of-the-valley, see *Convallaria majalis*
Lipotropic factor treatment, 19
Liverworts, 172
Livestock, *see* Domestic animals and livestock
Lonkar, A., 174
LSD, 28, 29, 30, 31
Lupinus, 61
Lycopersicon, 59, 71; *L. esculentum,* 70
Lycopodium clavatum, 122
Lycorine, 118
Lymphocyte mitogens from plants, 83–102
Lymphocytes, 83, 84, 86, 87, 88–89, 90, 91, 99, 171; B cells, 86–87, 88–89; mouse spleen, 86, 87, 90–91, 92, 99; T cells, 86–87, 88–89, 91

Magnesium salts, 132
Malagasy Republic, 142
Mancineel tree (*Hippomane mancinella*), 140–41, 143
Mandrake, American (*Podophyllum pel-tatum*), 107, 120–21, 128
Mangifera indica (mango), 161, 171
Maranta leuconeura, 108, 127; *M. leu-coneura* var. *kerchoveana,* 104, 107, 119
Marigold (*Tagetes*), 173, 177, 178, 179
Massachusetts Institute of Technology, 89
Mast cells, 86
Maya, 28

Mayapple (*Podophyllum peltatum*), 107, 120–21, 128

Mechanism of toxic activity (biochemical): abrin, 93; amatoxins, 14–16; bacterial toxins, 93; contact allergens, 171; *Coprinus*–ethanol reaction, 43; *Dieffenbachia* species, 123–24; 1,25-dihydroxyvitamin D_3, 63; furanocoumarins, 178; gyromitrin and analogs, 38; hypoglycin A, 43; mitogen binding to lymphocytes, 85–86; muscimol, 27; phallolysin, 21; phallotoxins, 20

Medicinal preparations (general), 69, 137

Meperidine, 132

Metabolism studies: acetaldehyde, 41; alkylcatechols, 166; *Amanita muscaria* constituents, 22; amatoxins, 16–17; coprine, 43; gyromitrin and analogs, 37; hypoglycin A, 43; ibotenic acid, 26; nicotine, 66; psilocin, 29–30, 31

Metal ions, 84

Methyl-α-amanitin, 16

Methyl-demethyl-γ-amanitin, ³H, 17

N-Methyl-N-formylhydrazine, 36–37

N-Methylhydrazine, 36–37

β-(Methylenecyclopropyl)-alanine, 43

Metopium toxiferum, 161; urushiol, 164–65

Mexican marigold (*Tagetes minuta*), 177, 179

Mexico, 28, 29, 177

Microorganisms, 83

Microscopy: electron, 85, 96, 97; UV, 85

Milk, 132

Milkweed, 61

Mitogens, lymphocyte, 83–102

Mole deterrents, 148

Monstera, 108; *M. deliciosa*, 104, 107, 119, 127

Moraceae, 178

Morels: false (*Gyromitra, Helvella, Verpa*), 35; true (*Morchella*), 35

Morphine, 64

Mosquito repellents, 116, 174

Mules, 64

Muscarine, 8, 23, 32–35; biosynthesis, 33; stereoisomers, 33; structural studies, 33

Muscimol, 8, 21–27, 33

Muscular toxic effects due to plant ingestion, 22, 25, 26, 72, 131

Mushroom genera: *Amanita*, 8, 9, 13, 19; *Chlorophyllum*, 8; *Clitocybe*, 8, 33, 34; *Copelandia*, 29; *Coprinus*, 40, 41, 43, 44; *Cortinarius*, 8; *Entoloma*, 8; *Galerina*, 8, 10, 32; *Gymnopilus*, 32; *Gyromitra*, 35; *Helvella*, 35; *Hypholoma*, 8; *Inocybe*, 8, 33, 34; *Lepiota*, 8, 10; *Morchella*, 35; *Panaeolus*, 29, 32; *Psilocybe*, 29, 30; *Stropharia*, 29; *Tricholoma*, 8; *Verpa*, 35

Mushrooms (general): collection for food, 8; fatalities due to, 7, 9, 10; hospitalization due to, 7, 8; ingestion by children, 7; "little brown," 32; poisoning by, 7

Myoblasts, 83

National Center for Poisoning Information, 130

National Clearinghouse for Poison Control Centers, 3, 7, 103, 130

Nephrolepis, 133; *N. exaltata*, 105, 107, 108, 119, 127

Nerium, 104, 108; *N. oleander*, 107, 119, 127, 131

New Testament, 22

Nicotiana, 59; *N. tabacum*, 60, 65–69

Nigeria, 145

Night-blooming cactus, see *Datura stramonium*

Nightshade family, see Solanaceae

Norbaeocystin, 30

North America, 9, 10, 29, 161, 171; see also United States

Nubians, 64

Octaeicosanol, 115

Oleander (*Nerium oleander*), 107, 119, 127, 131

Oncogens in plants, 39, 40, 154

Oral antiseptic agent, 145

Orfila, M. J. B., 3

Organotherapy, 19

Ornamental plants, 4–5, 103–35, 148, 150, 154

Ovum, 83

Paeonia, 120, 133; *P. lactiflora*, 105, 107, 108, 120, 127

Panaeolus cyanescens, 31

Papilionaceae, 178

Parasympathetic nervous system, toxic effects on due to plant ingestion, 24, 32, 63, 131
Parthenium hysterophorus, 174–77
Patch tests, 171, 173, 175
PCP, 31
Peanuts, 60
Pencil tree, see *Euphorbia tirucalli*
Pentadecyl: -catechols, 165, 166, 168; phenol, 165; -resorcinols, 165; -veratrole, 166
Peperomia, 127, 133; *P. leprostachya*, 120; *P. obtusifolia*, 104, 107, 108, 120, 127
Peritoneal dialysis, 18
Persian insect powder, 174
Phallotoxins, 19–21; seco derivatives, 19; thio derivatives, 19
Pharmacological effects of poisonous plant constituents, see Toxicity, symptoms
Phaseolus vulgaris, 84
Phencyclidine, 31
Phenols, alkylated dihydroxy, 165, 166, 168
Phenylbutazone, 17
Philodendron, 104, 108, 132; *P. cordatum*, 120; *P. oxycardium*, 107, 120, 127; *P. sagittifolium*, 105, 107, 120, 127
Philodendron, cut-leaf, see *Monstera deliciosa*
Phorbol derivatives: A-factors, 138; B-factors, 138; deoxyhydroxyphorbols, 139, 140, 141, 149, 152; deoxyphorbols, 139, 140, 149, 152; 12-deoxyphorbol esters, structural requirements for irritancy, 152–53; diesters, 138, 139; $\Delta^{1,10}$-isophorbol, 141; mancinellin, 140; phorbol, 116, 137, 138, 141, 149, 151; phorbol-12, 13-didecanoate, 152; semisynthetic esters, 152; TPA (12-O-tetradecanoylphorbol-13-acetate), 84, 138, 151, 152; triesters, 138, 139
Photodermatitis, see Dermatitis, photo-
Physicians, 4, 7
Physostigmine, 131
Phytohemagglutinin (PHA), 84, 86, 90, 99
Phytolacca americana, 84, 87
Phytophthora infestans, 75
Phytotoxicology, problems of, 1–6, 104, 129–30
Pilea cadierei, 104, 107, 108, 120, 127–28
Pilzatropine, 24
Pisum sativum, 84

Plants, household, see Ornamental plants
Podophyllum, 104, 108; *P. peltatum*, 107, 120–21, 128
Poinsettia, see *Euphorbia pulcherrima*
Poisindex, 130
Poison control centers, 3, 4, 7, 115
Poison hemlock (*Conium maculatum*), 68
Poison ivy (*Toxicodendron radicans*), 161, 164–65, 171
Poison oak: eastern (*Toxicodendron toxicarium*), 162; western (*T. diversilobum*), 162, 164–65, 171
Poisonous plants: fatalities due to, 4; hospitalization due to, 4, 103; identification of, 3–4, 5; labeling, 4–5; literature of, 1, 3, 4, 5–6
Poison sumac (*Toxicodendron vernix*), 162, 164–65, 171
Poison wood tree (*Metopium toxiferum*), 161, 164–65
Pokeweed (*Phytolacca americana*), 84, 87; mitogen (PWM), 84, 87–89, 90
Poland, 35
Polyacetylenes, 179–80
Polysaccharides, 83
Port of Spain, Trinidad, 177
Potassium salts, 132
Potato, see *Solanum tuberosum*
Prayer plant, see *Maranta leuconeura*
Primary plant constituents, 1
Propranolol, 132
Proresiniferatoxin, 144
Proteins in plants, poisonous, 89, 92–96, 113
Psilocin, 8, 28–32
Psilocybe: *P. baeocysis*, 30; *P. mexicana*, 29, 30; *P. pelliculosa*, 31; *P. semilanceata*, 31
Psilocybin, 8, 28–32; biosynthesis, 30; stability, 29
Psoralens, 178
Pupillary dilation, 35, 63
Purgatives, 108, 109, 110, 112, 137, 148
Pyrethrin II, 177
Pyrethrum, 174
Pyridine/piperidine alkaloids, 59, 65–69; structural requirements for teratogenicity, 69
Pyridoxal phosphate, 38
Pyridoxine hydrochloride, 38

Ragweed (*Ambrosia*), 173, 174, 176, 178

Rangelands, 60

Recreational use: *Amanita muscaria*, 23; psilocybin-containing mushrooms, 31; wild plants, 103

Registry of Toxic Effects of Chemical Substances, 2

Renal failure due to plant ingestion, 10, 16, 21, 35, 120, 131

Resin, 114, 115, 116, 176; *Dracaena*, 114; podophyllin, 121, 128

Respiratory organs, toxic effects on due to plant ingestion, 32, 35, 40, 61, 72, 131

Rhubarb substitute, 110

Rhus vernicifera, see *Toxicodendron verniciferum*

Rig Veda, 22

RNA, 14, 15

Robinia lectin, 90, 98–100; purification, 98; structural studies, 99–100

Robinia pseudoacacia, 89, 98–100

Rocky Mountain Poison Center, Colorado, 7

Rosaceae, 131

Rosary pea, see *Abrus precatorius*

Rubber, 114, 116, 137, 145

Rubber plant, see *Ficus elastica* (*decora*)

Rutaceae, 178

Saintpaulia, 121, 128; *S. ionantha*, 105, 107, 108, 121

Saline laxatives, 131, 132

Sandbox tree, see *Hura crepitans*

Sansevieria, 108; *S. thyrsiflora*, 104, 107, 121; *S. trifaciata*, 121, 128; *S. trifaciata* var. *laurentii*, 104, 107, 121

Sapium japonicum, 139

Saponins, 104, 114, 121, 128, 129

Scabiecidal agent, 174

Secondary plant constituents, 1–2, 4

Selaginella: *S. pallescens*, 104, 108, 121, 128; *S. rupestris*, 122

Semecarpus, 171

Serotonin, 30

Sesquiterpene lactones, 171–74, 175, 176, 177, 180; biological activity of, 173; Michael-type addition with sulfhydryl groups, 176

Sex hormone therapy, 19

Shaggy mane (*Coprinus comatus*), 40

Sheep, 64, 72

Skeletal toxic effects due to plant ingestion, see Calcinosis due to plant ingestion

Skin: blistering agents, 121, 148; carcinogenesis (chemical), 137, 148, 150–51; disease treatments, 120, 124, 127; hyperplasia, 151; irritants, 108, 109, 110, 111, 113, 115, 116, 117, 119, 120, 122, 123, 127, 137–59, 168, 177; toxic effects on by plants, see Dermatitis

Snake plant (*Sansevieria trifaciata*), 121, 128

Sodium EDTA, 132

Sodium metaperiodate, 84

Solanaceae, 6, 59–82, 104, 108, 128, 131; see also names of genera

Solanum, 59, 71, 104, 108; *S. malacoxylon*, 61–63; *S. melongena*, 70, 75; *S. pseudocapsicum*, 107, 122, 128, 131; *S. tuberosum*, 60, 70, 74, 75, 77–79

Solanum alkaloids, 69–72, 74–79, 122, 131; structural requirements for teratogenicity, 75–79

Soma, 22

South Africa, 140

South America, 29, 62

Soybeans, 60

Sperm, 83

Spurge family, see Euphorbiaceae

Steroids, 104

Sterols, 124; β-sitosterol, 115; tirucallol, 114

Stropharia cubensis, 31

Sudatory agent, 109, 111

Sugar beet, 60

Sugars, 83

Sunlight and dermatitis, 178

Sweden, 42

Switzerland, 24

Syngonium, 128; *S. podophyllum*, 108; *S. podophyllum* var. *albo-virens*, 104, 107, 122

Tagetes: *T. minuta*, 177, 179; *T. patula*, 178

Tannin, 111

Tansy, 173

Tapioca (*Manihot*), 145

Teratogenic effects due to plant ingestion, 59, 61, 65, 67–79

Tetracaine, 132

Thiophene dervatives: α-terthienyl, 178–79,
    180; 5-(3-buten-l-ynyl)-2,2′-bithienyl, 178–
    79
Thymidine, ³H-, 86, 90, 99
Tigliane, 139; orthoesters, *see* Daphnane es-
    ters
Tobacco, see *Nicotiana tabacum*
Toxicity: definition, 1; general, of plants with
    known toxic constituents, 5; problems of
    study of, 6; symptoms: acetaldehyde, 41;
    Amanita muscaria, 22; A. *phalloides*,
    10–11; amatoxins, 10–11; Anacardiaceae
    species, 161; *Arisaemia triphyllum*, 108–9;
    atropine, 35; *Begonia* species, 110; *Blighia
    sapida*, 43; *Caladium bicolor*, 106, 109,
    111; *Cestrum diurnum*, 61–62; C. *parqui*,
    61; *Codiaeum variegatum*, 111; *Coprinus
    atramentarius* and ethanol, 40, 41; *Con-
    vallaria majalis*, 106, 109, 112; *Datura
    stramonium*, 63, 64, 65; *Dieffenbachia*
    species, 106, 109, 112–13, 123–24; *Dig-
    italis purpurea*, 106, 114, 124; *Dracaena
    cylindrica*, 114; *Euphorbia* species, 106,
    114–16, 124–26, 132, 153, 154; *Ficus
    carica*, 116; *Gyromitra esculenta*, 35,
    37–38; *Hippeastrum* species, 106, 118,
    126; *Lantana camara*, 106, 118–19,
    126–27; lectins, 83; *Monstera deliciosa*,
    107, 118, 127; muscarine-containing
    mushrooms, 24, 32; muscimol and
    ibotenic acid, 25–27; *Nerium oleander*,
    107, 119, 127; *Nicotiana tabacum*, 65–
    69; nicotine, 66; *Parthenium hysterophorus*,
    174–77; phallolysin, 21; phallotoxins, 19–
    20; *Philodendron* spp., 107, 120, 127;
    Podophyllum peltatum, 107, 121, 128;
    psilocybin and psilocin-containing mush-
    rooms, 25–27, 28; *Solanum malacoxylon*,
    62–63; S. *pseudocapsicum*, 107; steroidal
    alkaloids, 72–79, 131; *Xanthosoma* spp.,
    122, 128
Toxicodendron: T. *diversilobum*, 161, 171; T.
    radicans, 161, 171; T. *toxicarium*, 162; T.
    verniciferum, 163, 171; T. *vernix*, 162,
    171; urushiol, 162–68 *passim*
Toxin, definiticn, 161
Treatment of plant poisoning: *Amanita
    muscaria*, 27; amatoxins, 10, 16, 17–19;
    aroids, 132; cardiac glycosides, 131–32;

cyanogenic glycosides (cyanide poisoning),
    131; *Euphorbia* spp., 132, 154; general, 4,
    130–33 gyromitrin and analogs, 38–39;
    muscarine, 34–35; poisonous Anacardi-
    aceae, 168–69; psilocybin-containing mush-
    rooms, 31; solanum alkaloids, 131
Trinidad, 177
Triterpenes, 115, 118, 119, 124, 127
Tropane alkaloids, 24, 27, 34, 59, 63–65,
    131, 132; atropine, 24, 27, 34, 63, 64, 65,
    118, 131, 132; hyoscyamine, 24, 63;
    scopolamine, 24, 27
Tropine, 63
Tumor-promoters, 137–59
Tumor-promotion, 137, 151, 154
Tung fruit (*Aleurites fordii*), 140, 148
Tung oil, 145, 148

Ultraviolet light, 178
United Kingdom, 24, 145, 148
United States, 7, 9, 10, 11, 18, 31, 34, 35,
    40, 42, 44, 60, 63, 65, 103, 145, 148,
    153, 154, 161, 162, 175; East, 9, 35, 64,
    162; Hawaii, 124; Midwest, 35, 64, 67;
    South, 31, 61, 62, 64, 126, 161, 162, 174,
    175, 177; Southwest, 61, 63, 175; West, 7,
    8, 9, 27, 31, 35, 72, 130, 172
United States Department of Agriculture, 60
University of British Columbia, 171
Urinary tract, toxic effects on due to plant
    ingestion, 10
U.S.S.R., 22

Venereal disease treatments, 116, 121, 122,
    128, 148
Veratrum alkaloids, 69–75, 76, 77; structural
    requirements for teratogenicity, 73–74
Veratrum, 59, 61; V. *californicum*, 72
Vertebrates, 1, 5
Vesicants, 121, 148
Vicia faba, 84
Viruses, 83
Visual disturbances due to plant ingestion or
    exposure, 22, 24, 25, 28, 32, 63, 114,
    116, 121, 125–26, 132, 154
Vitamin B₆, 38
Vitamin D₃, 62
Vitamin therapy, 19

Vocal disturbances due to plant ingestion, 109

Wart treatments, 148
Water, 132
Watermelon plant, see *Pilea cadierei*
West Germany, 173
West Indies, 43, 109, 161

Wheat, 60
Wisteria floribunda, 84
Wound healing treatment, 116

Xanthosoma, 128, 132; *X. hoffmannii*, 105, 107, 108, 122
X-ray crystallographic techniques, 12, 33, 94, 96, 141